APPLIED MATHEMATICS

Applying Statistics and Probability to Everyday Life

Erik Richardson

Cavendish Square
New York

Published in 2017 by Cavendish Square Publishing, LLC
243 5th Avenue, Suite 136, New York, NY 10016

First Edition

CPSIA Compliance Information: Batch #CS16CSQ

All websites were available and accurate when this book was sent to press.

Library of Congress Cataloging-in-Publication Data

Names: Richardson, Erik, author.
Title: Applying statistics and probability to everyday life / Erik Richardson.
Description: New York : Cavendish Square, 2017. | Series: Applied mathematics | Includes index.
Identifiers: ISBN 9781502619730 (library bound) | ISBN 9781502619747 (ebook)
Subjects: LCSH: Mathematical statistics--Juvenile literature. | Probabilities--Juvenile literature.
Classification: LCC QA276.13 R534 2017 | DDC 519.5--dc23

Editorial Director: David McNamara
Editor: B.J. Best
Copy Editor: Nathan Heidelberger
Art Director: Jeffrey Talbot
Senior Designer: Amy Greenan
Production Assistant: Karol Szymczuk
Photo Researcher: J8 Media

The photographs in this book are used by permission and through the courtesy of: Peter Ginter/Getty Images, cover; Filip Lolić/File:M45 filip.jpg/Wikimedia Commons, 4; PhotoAlto/Laurence Mouton/Getty Images, 8; Christian Albrecht Jensen/File:Carl Friedrich Gauss 1840 by Jensen.jpg From Wikimedia Commons, 11; NASA, ESA and E. Karkoschka (University of Arizona)/File:Saturn HST 2004-03-22.jpg/Wikimedia Commons, 14; Alison Wright/Corbis Documentary/Getty Images, 16; Francis Galton/File:Galton's correlation diagram 1875.jpg/Wikimedia Commons, 19; lzf / Shutterstock.com, 20; goodluz/Shutterstock.com, 22; Victor De Schwanberg/SPL/Getty Images, 32; Another Believer/File:Licenciado Gustavo Díaz Ordaz International Airport (2014) - 03.JPG/Wikimedia Commons, 49; Rubberball/BrandX Pictures/Getty Images, 50; Ahodges7/File:Manatee at Sea World Orlando Mar 10.JPG/Wikimedia Commons, 63; Ian Sutton (http://flickr.com/photos/22616984@N07)/ File:Parkes Observatory -Australia-15Aug2009.jpg/Wikimedia Commons, 74; Bill Greene/The Boston Globe/Getty Images, 76; NOAA, 100; Drnathanfurious at English Wikipedia/File:Benford-physical.svg/Wikimedia Commons, 114; Jag_cz/iStock/Thinkstock, 117.

Printed in the United States of America

TABLE OF CONTENTS

We tend to notice the stars, but when you think about it, our universe is mostly empty space in between stars.

INTRODUCTION

It was an amazing shift in the world and our ability to make sense of it when, in 1926, Erwin Schrödinger showed us that the atom did not work by means of little **electrons** following a set orbital path. In fact, it is a probability function. For the first time we realized on even the atomic level that we did not live in a certain world that runs on grooved tracks. Rather, we live in a world that has probability, uncertainty, and things that might or might not be there.

That last bit might sound strange, but let's think about it for a minute. Most of our universe is emptiness with some stars sprinkled here and there like golf balls adrift in an ocean. Most of an atom is empty space, too. A common model that will help you get your head around this: if the nucleus of an atom were the size of a pea, and you set it down on the 50-yard line of a football field, the electron cloud would be out near the bleachers. That's a lot of empty space inside stuff we think is solid.

At the same time, that emptiness makes our world a paradox of such awesomeness that it makes your brain spin a little to think about it. Consider how the "average student" doesn't actually exist. Everyone is off the **mean** in one direction or another. Yet how much of the students' and teachers' lives are built around teaching that invisible "average student," making him feel like he belongs? We build desks so he will be comfortable. We chop lessons to fit his attention span. And on and on.

That's just a start. Think about the fact that there are imaginary numbers (like i, $\sqrt{-1}$). That number doesn't even exist, but when we plug it into our equations, it makes sure our airplane designs stay in the air and our cars stay on the ground. Don't even get me started talking about infinity, though there is a little infinity surprise hiding later in the book.

So what are we up to? Where are we going in this book? Well, I propose that the model we put in our heads is that of the explorers. Statistics is very much like a compass or **altimeter** or a sonar system—it is a diverse set of tools that allows us to navigate in regions where we can't see very well or far. The world is a dense, vibrant rain forest. It is an ocean that we have only ever explored 10 percent of, and it is space that we have barely touched in our explorations. Statistics can help us find the scenic lookouts and avoid the poison dart frogs, to delight in the coral reefs and avoid becoming a Kraken snack. We want to head out into the world because there is mystery and discovery (you can't have one without the other), but we want to find the dazzling **supernovae** and avoid getting sucked into a black hole. Good navigation instruments will help us to do exactly that. They help us filter out what is random noise and what is important information.

ONE

The History of Statistics and Probability

To help ourselves navigate into unknown or unfamiliar areas, it can help to frame our exploration in terms of models that allow us to measure our terrain and our progress. Then we can adapt in response to those measurements, repeating this process of measuring and adapting as we continue onward.

While statistics and probability are very closely intertwined, understanding how they arose separately can provide context for working with them in contemporary settings. Statistics in a broader sense arose earlier, though it only really evolved into what we now think of as statistics after it came into closer interaction with the math that developed out of work in probability theory. Because of that, let's talk about probability first.

Born in a Casino

Well, OK, not quite a casino as we might think of it these days, with slot machines and all that, but in gaming rooms and gambling houses. Arising out of a discussion

Sometimes we come up against puzzles about numbers but aren't able to tackle them until a new kind of math develops.

of figuring out the stakes and bets on certain games, two super-smart mathematicians picked the problem up and discussed the ideas in a series of letters back and forth. The two mathematicians were Pierre de Fermat and Blaise Pascal, and the year was 1654.

While they came up with different ways to approach the problem, their dialogue launched the foundations of probability theory. An important step forward came in 1657 when the Dutch thinker Christiaan Huygens published a short text on the subject, giving the topic its first scientific treatment. More importantly, it opened up ideas to show that the importance of probability extended beyond games of chance to include other contexts like fair legal contracts and proper evaluations of risk in investments. This led to the publication of a short book about probability and annuities by Jan de Wit, the leader of the Netherlands. (Annuities are a kind of investment that were often used by governments to raise money—especially for stuff like wars.)

Pascal would go on to develop the idea that probability could also come into play in such grand pursuits as the attempt to wrap our heads around ideas about God and the afterlife. He developed

Christiaan Huygens

Christiaan Huygens was a brilliant Dutch scientist and mathematician who lived from 1629 to 1695. In studying so many great figures of the past, it is hard to deny that math seems to do something to expand the mind.

Huygens was homeschooled until the age of sixteen, and Rene Descartes was impressed by the boy's talent for geometry even when Huygens was in his teens. When he went off to college, he took up corresponding with Marin Mersenne, who also wrote about Huygens's ability in the field of math (even comparing him to legendary Greek mathematician and scientist Archimedes).

Huygens would go on to develop his talents after college, and he grew to become one of the most important scientists between Descartes and Newton. One of his more notable achievements was inventing the pendulum clock (along with interesting mathematical proofs related to the way pendulums behave). He was also the teacher and math tutor of Gottfried Leibniz—one of the two guys who invented calculus. Huygens also came up with a method for estimating the distance of stars, and he used a telescope to discover the rings of Saturn as well as its largest moon, Titan.

Among the things named after Huygens in tribute are a crater on Mars, a mountain on the moon, and the probe that landed on Saturn's moon Titan in 2005.

one of the most famous arguments about living a good life in his book *Pensées*.

As we move forward, we find Gottfried Leibniz writing about the idea, in 1677, of a world where most disagreements and conflicts could be worked through by means of reasonable calculations of probability. (Some of you might recognize Leibniz as one of the super-geniuses who, among other things, invented calculus.)

The year 1713 brought us Jakob Bernoulli's development of the law of large numbers. This helped to explain how outcomes would converge toward their natural probability as the number of trials increased. (The quick tour here is that a series of six coin flips often does not come out 50/50, but a series of one thousand coin flips would get closer and closer.)

There were a few more steps along the way involving some technical concepts in statistics, but the next one worth touching on here is the use of probability in the 1750s and 1760s as part of a debate about vaccinating for smallpox.

The year 1761 brings us to Thomas Bayes and his work on conditional probability. Conditional probability explores what the chances are that B will happen if A has happened. The next year, 1762, the first insurance policies incorporating probability started being written.

Pierre-Simon Laplace in 1774 gave the first push to figure out a rule for combining observations based on the theory of probabilities, and by 1778 he had developed the idea of the normal distribution. A normal distribution suggests that all probabilities of an event can be graphed into a shape that looks like a bell (hence the name "bell curve"). This idea was later redeveloped by Carl Friedrich Gauss and ended up with his name on it, rather than Laplace's.

In order to prepare for the point where probability and statistics get married and begin the rise toward our current approach, let us go back and follow the line of development for statistics.

Carl Friedrich Gauss helped develop the modern normal distribution curve.

Ancient Beginnings

Some basic versions or partial versions of statistics have been in play since practically the dawn of time. We should think of it more as **data** collection than our modern idea of "statistics." Early nations like Rome often used various kinds of data collection and compilation—like putting together a type of census to record the trading in various **commodities**.

There is an example of using data to solve a puzzle included in Thucydides's book *History of the Peloponnesian War* from the fifth century BCE. In the story, we see that the Athenians figured out the height of a certain wall for a city they were attacking by counting bricks in a section of the wall where the plaster had come off. Different people counted, and those counts were double-checked by other soldiers. The number that was given the largest number of times (what we call the mode) was considered to be the best answer. The Athenians multiplied that number by the height of one brick to figure out how tall their ladders needed to be in order to successfully scale the walls and take over the city.

In the *Mahabharata*, the classical Indian epic, a certain king estimated the number of leaves and the fruit on a given tree (2,095 fruit and 50 million leaves) by counting them on one twig. Then he multiplied by the number of twigs on the branches. His estimate turned out to be very close to the actual total.

These are two early examples of starting from a small sample and generalizing to a larger group—the identifying feature of statistics (as opposed to probability).

Another valuable example of something that would come to be a staple of modern statistics appears in a ninth-century book by the Arabic thinker Al-Kindi (801-873 CE). The book was named *Manuscript on Deciphering Cryptographic Messages* (well, except, you know, in Arabic, not English), and it gave careful instructions on how to use statistics and **frequency** analysis to help untangle coded messages. This book helped establish the fields of statistics and **cryptanalysis**.

Through the Middle Ages and Beyond

As we move into the medieval period, we find a clever early example of using statistical sampling for quality control. This process was devised to help make sure that silver coins from one batch to the next still contained the same amount of silver. Dating back to the twelfth century, each time the royal mint turned out a batch of new coins, one of them was placed in a locked box in Westminster. The box was call the Pyx, so the procedure was referred to that the Trial of the Pyx. After a certain amount of time, the box was opened and the coins in there were weighed against the new batch to check for consistency of weight. If the coins weighed the same, it meant they had the same amount of silver.

Then, in the fourteenth century, a banker in Florence, Giovanni Villani, produced the *Nuova Cronica*, which is a history of Florence. In it, though, he included a lot of statistical information—population, commerce, ordinances, and more. The book is considered an important stepping stone in developing modern statistics.

It would be almost two hundred years before the next big step, as the field broadened from not just the collecting such data but toward summarizing and analyzing it as well. During this period, the fields of probability and statistics became more intertwined. By the late 1500s, statistics had come to rely on probability as the heart of this systematic approach. This synthesis allowed the field to expand from history and economics, which were just scratching the surface. Now statistics was able to be used in fields like experimental psychology.

Our next step was calculating the arithmetic mean (commonly known as the average). Although it was a concept known to the Greeks, there was no formula to calculate the mean for more than two values until the sixteenth century. This method was based on work done in astronomy by Tycho Brahe as he worked to fix the positions of various **celestial** bodies.

The next stop on our trolley is 1599. Here we find the development of the **median** as a useful measure of the central value of a set of data. The median is literally the middle number in the set. This idea is credited to Edward Wright's book about navigation, entitled *Certaine Errors in Navigation*. He was discussing figuring out a location with a compass.

Maxwell and Boltzmann used statistics to analyze the speed and energy of gas particles. Their work helped explain why filling a balloon with heated air can make it fly.

The field of statistics was hot

By the mid-1800s, scientists had been studying the physics of heat for a long time. Though it may be hard for us to believe now, there was still a lot of disagreement about whether all matter is made up of tiny particles—atoms.

James Clerk Maxwell, the same thinker who figured out how to show that magnetism and electricity are related, developed a law to explain how gas particles behave. Since there are too many molecules bouncing around to measure the speed of each, his idea (1859) was to use statistics and show that the speeds are spread out in a distribution (kind of the way student scores on a test are spread out—some high, some middle, some low).

Maxwell's work sparked the thinking of Ludwig Boltzmann, who was a student at the time. Building on Maxwell's work, Boltzmann came up with an even better version of Maxwell's law (1871) that used the distribution of energy rather than just a distribution of speeds. This model came to be known as the Maxwell-Boltzmann distribution. Among other things, it was the only theory able to explain the function of the newly developed hot-air balloon. In particular, it was able to account for the way heating the air in the balloon caused the volume to increase while the density decreased.

This is Westminster Abbey, where the box for collections became a tool for random sampling much like that in statistics.

John Graunt, an English statistician, published a collection of numbers and data about London's **mortality rates** in 1662. His work was based on information and charts that had been compiled to track the rise and fall of the Plague. Graunt teamed up with William Petty to pursue their shared interest in **demography**. *Natural and Political Observations Made upon the Bills of Mortality*, the name of his book, produced the first life table (giving probabilities of living to each age) after he had sat down and analyzed the death rates in London. This work allowed Graunt to make the first statistically based estimation of a city's population. He also used funeral numbers (13,000 per year) and parish records to figure out that mean family size in London was eight and that the population was at about 384,000. He is fairly considered, because of this work, as a father of modern statistics.

In the 1680s, William Petty expanded this type of work in a series of essays on what he called "political arithmetic." His essays combined data with some attempts to calculate things like the economic value of all the people living in Ireland. This work by Graunt and Petty helped fuel developments in the use of statistics as a tool of government well into the eighteenth century.

Probability and Statistics in the Seventeenth Century and On

The year 1786 brought William Playfair's examples of graphical representation of statistics. Playfair is credited with inventing the line chart, the bar chart, and the histogram, having incorporated them into his work on economics, the *Commercial and Political Atlas*.

In 1795, Playfair invented the pie chart and circle chart. He used them to visually represent how England's imports and exports changed over time. These charts became widely known when he published his *Statistical Breviary* in 1801.

The method of least squares, which was used to minimize errors of statistical analysis, was published independently by three different scholars, of whom Gauss is most widely known. In 1801, Gauss used this method to predict the location of the dwarf planet Ceres, which lies between Mars and Jupiter.

Adolphe Quetelet (1796–1874), though you may not have heard of him, invented the notion of the "average man." This idea is a tool to help understanding complex social **phenomena** such as marriage rates, crime rates, and suicide rates.

Finally, in the 1870s, we get the first tests of the normal distribution—the bell curve. These were invented by the German statistician Wilhelm Lexis. He was able to show birth rates were normally distributed.

The original scope of statistics was limited to data useful for government. During the nineteenth century, this approach was stretched out to accommodate other commercial and social fields of inquiry. This framework was based in the similarly growing probability theory.

Developments in Modern Statistics

While the theory under statistics is rooted in eighteenth-century probability, the modernization of the field unfolded in several stages. Through the nineteenth and twentieth centuries, statistics became more mathematically rigorous—not only for science but for industry and politics, too. The next shift was the development of better experimental design, and techniques to use with small data samples. These developments then became further refined and improved.

With the more recent developments, the most valuable single name to know is Francis Galton, who is credited as one of the principal founders of statistical theory. His contributions to the field include the concepts of **standard deviation**, **correlation**, and **regression**. He applied these methods to study human physical characteristics, such as height and weight.

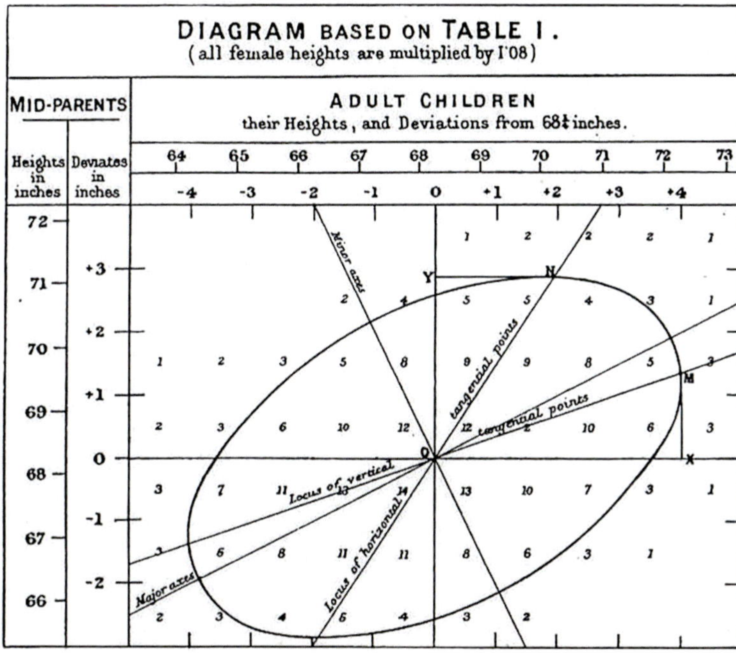

This chart from Francis Galton in 1875 shows the relationships between the heights of parents and their adult children.

In 1907, Galton published a paper in *Nature* where he described how useful the median could be. He analyzed 787 guesses of the weight of an ox at a country fair. The actual weight was 1,208 pounds; the median guess was 1,198. The guesses were very scattered, but the median was remarkably close. This meant the median was a useful tool in some situations.

This brief history should give us a working sketch of the range of instruments and their origins. Now let's head out on our expedition to see new sights and to see how we are able to navigate in the different environments with the help of these instruments and gauges!

Statistics can be thought of as a tool to help us navigate, much like a compass for hiking away from the familiar paths.

TWO

Statistics and Probability in Your Everyday Life

Here we will begin the first part of our voyage. In this chapter, we will start to familiarize ourselves with some of the dials and gauges and radar systems that allow us to navigate through the forests and fog of our daily lives. By looking at some of the different places we visit, we will also see different examples of how the instruments of statistics and probability allow us to bring things into view that might have been hidden, or to bring into focus things that were blurry. Among the scenic lookouts, so to speak, we will go from sitting in a student desk to standing on Wall Street. Along the way, we will eat breakfast, go for a drive, vote for president, and several other things. Each time, we will understand more and more about how the instruments work and what they mean, as long as we're paying attention and if we keep our brains turned on and thinking as we travel.

If you concentrate, this journey will also stretch your mind as we go by, giving you ideas you haven't tried to wrap your head around before. There's no succeeding or failing if you don't understand some of these completely. These instruments are cool gadgets, and that means they're fun to play and tinker with. Take that mindset into the process.

Student

The first place to look out from is right where you're probably standing (or, you know, sitting, since not many of us read standing up). The tools and instruments of statistics can help you make sense out of a much larger variety of questions and puzzles at school. Keep in mind that our goal here is not to walk away knowing how to do all this stuff. This is more like sampling to see what kinds of things we could do.

Guess what? You shouldn't guess

In her statistics class, Madisen has a quiz with four questions. Each is multiple choice with five choices. In order to pass, she needs to get at least three out of four questions right. She didn't study and has no idea of the right answers, so she is blindly guessing her way through the quiz. What is the probability that she manages to pass? What about the probability of getting them all right?

First we need the probability for each possible correct answer. Since there are five choices, her chances of being correct on any given question are:

$$p = \frac{1}{5} = 0.2$$

The equation for getting *x* correct questions is:

$$P(x) = \frac{n!}{x!(n-x)!}(0.2)^x(1-0.2)^{n-x}$$

where:

n = The number of questions.
x = The number of correct answers.

Remember that the ! symbol means "factorial." A factorial is a product of all integers from one up to and including the given number. So here, 4! = 1 × 2 × 3 × 4 = 24. So, the probability Madisen will get three questions correct is:

$$P(x) = \frac{4!}{3!(4-3)!}(0.2)^3(1-0.2)^{4-3}$$

$$P(x) = \frac{24}{6 \times 1} \times 0.008 \times 0.8^1$$

$$P(x) = 4 \times 0.008 \times 0.8^1$$

$$P(x) = 0.0256$$

Similarly, the probability that she manages to get all four correct is:

$$P(x) = \frac{4!}{4!(4-4)!}(0.2)^4(1-0.2)^{4-4}$$

You need to know that 0! = 1:

$$P(x) = \frac{24}{24 \times 1} \times 0.016 \times 0.8^0$$

$$P(x) = 1 \times 0.0016 \times 1$$

$$P(x) = 0.0016$$

We like to speak of the "average student," but no student is completely average in every possible way.

Middles that aren't there

For any set of data, we often calculate the mean, another term for the average. And when we find the line of best fit through regression, the median is one of the points through which that line of best fit has to travel. But the median isn't always one of the data points. That means the line of best fit might be anchored around a spot that isn't even there.

That's just the beginning. Whenever we want a way to measure whether or not a result is likely to have happened by chance or not, we calculate that by using something called the standard deviation. The standard deviation is a mathematical calculation of how spread out the data is. It should mess with your mind a little bit to think about how we are measuring from an imaginary point that isn't really there. That would be like putting up signs on the highway telling how many miles it is until we reach some town that isn't there—it's just an empty field.

I will leave it to you to consider why textbooks and desks and stuff are built to fit the average student when she doesn't exist either.

So her chance of getting them all right is a 0.16 percent probability. Her chances of passing, though, include both the option where she gets three questions correct and the answer where she gets all four correct. That means:

$$P(x) = 0.0016 + 0.0256 = 0.0272$$

That's a 2.7 percent chance. Maybe Madisen will study more next time.

I can't weight to find my grade

In order to figure out what grade you need on the final exam, you must go through the process of calculating the weighted average. In this case, though, you don't need to find the final weighted average but one of the missing values that would result in a specific final average. Suppose the points are set up in the following way:

Quizzes	20%
Midterm Exam	30%
Participation	10%
Final Exam	40%

Let's say you want to end the course with an overall grade of 85 percent. So, we take the scores earned for each and work to find how many points are missing from the overall target average.

Quizzes	20%	x	70	=	14
Midterm Exam	30%	x	85	=	25.5
Participation	10%	x	100	=	10
Final Exam	40%	x	____	=	____
			Total	=	85

For the values to add up to 85, the final exam score × 40 percent must equal 35.5 (since 35.5 + 14 + 25.5 + 10 = 85).

$$35.5 = 0.4x$$

$$35.5/0.4 = x = 88.75 \text{ percent needed on the final exam}$$

Forecasting college admissions and success

Different universities have different formulas they have created by regression applied to the data of incoming students to their schools. These formulas help predict how well incoming students might perform. Some factor in **standardized test** scores, some focus only on high school grades, and some mix several different pieces into the process. Let's consider a scenario where the admissions committee is using the following equation to predict the performance of potential incoming students:

$$y = 0.53 + 0.71x$$

where:

y = First-semester college grade point average (GPA).
x = High school GPA.

How would this forecast shift between students with a 3.0, 3.5, and 4.0 GPA in high school?

$$y = 0.53 + 0.71(4.0) = 3.37 \text{ GPA for the (previously) 4.0 student.}$$

Meet Daniel Taylor

Please meet Daniel Taylor, a student finishing up his last year of high school and going through the college admissions process.

Glad you could offer your thoughts, Daniel. Can you talk to us a little about how you found statistics to be helpful as a student?

There are a lot of general things that apply across classes, like being able to calculate the score you need to get on tests or how much would be needed to bring your homework average up. Plus, it can be very useful in subjects besides math, such as doing population studies in biology.

Do you feel like there are some important statistics that colleges should use to look at incoming students that they aren't using? What would you use as a meaningful metric if you could change it?

The grades in honors and AP classes are extremely important, but I wish they would also use the number taken as a measure of how hard students were willing to push themselves.

What are the kinds of statistics you have been using to help you pick the best university? Do you feel like some of those are more important than others? What is a statistic you wish they would track and share?

A lot of the ones you would expect, like cost and financial aid packages given out. It would be nice if they broke the numbers down more to show certain programs the student is interested in.

I wish schools had more data on how their students in specific majors end up later in life.

Any advice you would give younger students to help them prepare for the college application/admission process you're going through?

Many students think of their classes not as a way to learn, but as a mark on their transcript that will help them get into college. I know that is the mindset I had until my senior year. However, the closer I got to having to face college, the more that kind of evolved. So now I keep reminding myself that the purpose of school is to be more awesome in real life, not just look more awesome on paper.

Thank you for sharing your thoughts with us, Daniel. Best of luck in your admissions and in college! (I do solemnly swear on Darwin's beard and Napier's bones I did not bribe him to say that last part!)

$$y = 0.53 + 0.71(3.5) =$$
3.01 GPA first semester for the 3.5 student.

$$y = 0.53 + 0.71(3.0) =$$
2.66 GPA first semester for the 3.0 student.

Imagine if a school only admits students who average in the seventy-fifth **percentile** and above based on grades and standardized test scores. What percentile on a standardized test do you think you would have to earn in order to land a spot in the freshman class if your GPA put you in the seventieth percentile of your graduating class?

Health and Nutrition

Your body is like the ultimate gadget in its own right, so it's funny to think of using the gadgets and instruments of statistics to understand it. But then again, certainly a number of the gauges and screens on a sub or spaceship are things that monitor the ship itself to help maintain optimal conditions. I'll bet you probably never thought of your body as a submarine or a spaceship before, did you? See what I mean about having new ideas?

Anger danger

It's an interesting question to explore whether chronic heart disease (CHD) happens more often in people with high levels of anger. To see what the data suggests, let's look at the following chart:

	Low Anger	Medium	High Anger	***Totals***
CHD	54	111	27	*192*
No CHD	3,058	4,622	607	*8,287*
Totals	*3,112*	*4,733*	*634*	*8,479*

Let's imagine the researchers' conclusion is that anger and heart disease can be closely related. Does this data support that claim? A quick and simple way to test the data is just to run a rate comparison. The likelihood of CHD for the low anger group = 54/3112 = 1.7 percent; for the medium group = 111/4733 = 2.3 percent; and for the high anger group = 27/634 = 4.25 percent. That would suggest that yes, it seems that the higher the anger level, the more your probability of heart disease goes up.

Just a spoonful (or three or four) of sugar

Cereal	sugar (grams)
Frosted Mini Wheats	11
Raisin Bran	18
All Bran	5
Apple Jacks	14
Cap'n Crunch	12
Cheerios	1
Cinnamon Toast Crunch	10
Crackling Oat Bran	16
Fiber One	0
Frosted Flakes	12
Froot Loops	14
Honey Bunches of Oats	7
Honey Nut Cheerios	9
Life	6
Rice Krispies	3
Honey Smacks	15
Special K	4
Wheaties	4
Corn Flakes	3
Honeycomb	11

Sugar is like a math trick: it pushes the number of calories higher, but almost feels like it doesn't exist in terms of feeling full.

In looking at ways to make our eating habits better, it can help sometimes to characterize a whole group of foods. Consider, for instance, the list on page 31 of cereals and their sugar content per serving, based on information from the United States Department of Agriculture.

Given that list, we can calculate a couple of key numbers to help us in making a decision about whether to keep cereal in our diet or cut it as part of a smart overall nutrition plan. To find the mean for the sugar values, we just need to add them all together and divide by the total number of items on the list:

$$\text{mean} = \frac{11 + 18 + \cdots + 11}{20} = 8.75$$

Then to find the variance, we compare how far away each individual data point is from that average. This tells us, in a general way, how spread-out they are:

$$\text{variance} = \frac{(11 - 8.75)^2 + (11 - 8.75)^2 + \cdots + (11 - 8.75)^2}{19} = 28.30$$

If you take a second and look at what's happening there, we are measuring how far each actual value is away from the average. That's why you see 8.75 in each one—that's the average. The reason we square them is because that makes all the negatives into positives (negative times negative equals positive, right?). We just want their distances, not whether they are above or below. Then, to undo that positive-ification, we just take the square root of that value to get the standard deviation:

$$\text{Standard deviation is } \sqrt{28.30} = 5.32$$

The standard deviation is telling us the average distance away from the mean. It lets us figure out things like whether the scores are all bunched pretty close around the average or if they are really spread out.

Because things like human psychology and our modern time-crunched lifestyles make it hard to constantly evaluate each item we chomp, success with projects like creating healthy eating habits requires some management. One way to save time and gain an advantage over our own temptation, for instance, is to skip the cereal aisle altogether when we're zipping through the store. Statistics allowed us to draw some general conclusions about the group and see that, "for the most part," the amount of sugar in cereal would not match up with a lot of recommended nutritional targets.

Herbal remedies

One of the single most valuable things statistics does for us, day after day, is help to show which medicines and health procedures actually work. Statistical tests use objective evidence, careful evaluation of margins of error, and well-tested theories and models. The herbal medicine industry is one of the front lines in this battle. New supposed cures are announced almost weekly, and a variety of health claims are backed up by weak, flawed, or non-existent statistical legitimacy.

Here is a sample of a **randomized** study to see if a particular herbal treatment was more effective than a placebo when used to treat clients suffering with depression.

In the table below, subjects recorded their mood on the same rating scale. The study uses eight subjects, with four in the herbal treatment group and four in the placebo group. We see the results of their ranking on the depression scale after they underwent treatment for two months.

Herbal Treatment		Placebo	
Subject	Response	Subject	Response
2	44	1	89
4	95	3	60
6	78	5	69
8	70	7	50
Total	**287**	**Total**	**268**

Sample means:

$$\text{Herbal Treatment} = \frac{287}{4} = 71.75$$

$$\text{Placebo} = \frac{268}{4} = 67.00$$

We won't walk through the statistical calculations here. However, they would tell us that while the means look different, they are, in fact, *not* different enough to rule out random **fluctuation** of chance as causing the differences. Let me put that another way: because the sample size is small, and because the data bounces around a lot, we can't say one average is truly lower than the other.

Conversely, imagine two classrooms full of kids. For a morning snack, we'll give one class carrots and water. If we give the other class soda and bags of sugar, then we would expect to see most of the kids in that class bouncing around way more. They would have a much higher mean "activity level." (How to prove that the mean is actually higher is something that gets decided by crunching through the actual formulas and comparing to significance tables and stuff.) The sugar example is easy, though, because any of us who have ever had a snack like that, or seen a group do it, know that it is pretty obvious!

Religion

This is one of my favorites. This is the most mysterious territory of any that we could explore—even more than outer space. Wherever you stand on religious issues, statistics can help stretch your mind on the topic.

Partridge in a prayer tree

It's always interesting to see a gap between what people will say on a survey regarding their religious practices, and what they actually do. There are some times when this gap takes on a higher-than-average importance, like in the aftermath of the 9/11 terrorist attacks in the United States. A research group surveyed a random selection of 3,733 people and found that 991 of them said they had gone to a religious service in the days after the attacks on the World Trade Center and the Pentagon.

Researchers would naturally ask at this point what this number would mean with respect to the larger population. How would we calculate that with, say, a 95 percent **confidence interval**? A confidence interval means how likely it is that a measurement would be *x* standard deviations away from the average.

Take an example of a chapter test in social studies class. Let's say the class average was 80 (yes, we should have studied harder), and the standard deviation is 3 points. That means, on average, the students came within 3 points of 80 (so many of the scores are grouped in between 77 and 83). Now, if a student happened to have an 85, that wouldn't seem that impressive, right? It's just a little farther out than the main group. What about a student who scored a 42, though? He's way to the left when we make a chart, and he seems a long way from the average (and the rest of the group). Well, a confidence interval measures how rare or likely it would be for a score to land within a certain number of standard deviations without needing some special explanation. The student who scored a 42 is probably unusual; perhaps he felt very sick that day but took the test anyway.

For our religious service confidence interval, the idea works the same, but instead of how far a student score is likely to be from the group mean, we're talking about how far the mean of our sample group is likely to be from the mean we would get if we surveyed the entire population (like everyone after 9/11, not just the 3,733 people) and calculated that average score. To put it another way, what percentage of people *in the United States* attended a religious service?

Well, the sample size in the prayer study is 3,733, and the observed portion, meaning the group that said they attended a religious service, is:

$$p = \frac{991}{3733} = 0.2654 = 26.5 \text{ percent}$$

In order to get a confidence interval of 95 percent, we need a **variable** called $Z_{\alpha/2}$. We'd find this from a table, and here it would equal 1.96. We can calculate the margin of error E:

$$E = Z_{\alpha/2} \times \sqrt{\frac{p(1-p)}{n}}$$

$$E = 1.96 \times \sqrt{\frac{0.2654 \times 0.73453}{3733}}$$

$$E = 0.0141608 \approx 0.0142$$

To find the confidence interval, we take the percent of people in the sample who said they attended a religious service and then add/subtract the margin of error:

$$0.2654 \pm 0.0142$$

$$= 0.2512 \text{ lower bound}$$

$$= 0.2796 \text{ upper bound}$$

	Hours of Religious Activity (per month)		
Gender	*0*	*1–9*	*10–19*
Female	$\frac{231}{767} = 0.30$	$\frac{295}{767} = 0.38$	$\frac{90}{767} = 0.12$
Male	$\frac{276}{634} = 0.44$	$\frac{244}{634} = 0.38$	$\frac{58}{634} = 0.09$

The result is that we can be 95 percent certain that the percentage of the larger population who attended some form of religious service in the days following 9/11 is going to land between about 25 percent and 28 percent. This is how close the answer would be if we surveyed everyone instead of just surveying 3,733 people. It might be different than the answer we got, but it only has a 5 percent chance of being outside that range of 25 to 28 percent. Admit it, that's kind of cool that we can figure out something about everyone by just studying a small group of people!

Men's religious activity versus women's

Keep in mind that some key patterns and trends can be seen just from charting data and doing even one or two simple steps. We need not mobilize every tool for every statistics question we come across. Consider the example of a simple survey to find out whether men or women are more likely to spend time praying,

20–39	40+	Total	n
$\frac{101}{767}$ = 0.13	$\frac{50}{767}$ = 0.07	1.00	767
$\frac{42}{634}$ = 0.07	$\frac{14}{634}$ = 0.02	1.00	634

meditating, or reading books about religion and spirituality. A table like the one above is one way to get a quick overview and a general sense of whether it is more likely for men or for women. In each case, we have taken the data value for that cell and divided it by the total number of females (or males, depends on which row) to establish percentages. This will give you a quick intuition that allows you to go on and follow through with a more detailed analysis if it seems warranted.

From this, we can see that the genders are even at the level of one to nine hours each month. At that point, though, men's participation levels start to drop away to at a faster pace. Consider, for example, that only 7 percent of the men practice at the level of twenty to thirty-nine hours. Women are almost double that percentage. Once we look at the zone for forty hours or more each month, the men have dropped down to a third of the women.

Controlled studies for spiritual practices

Sometimes you will come across an article that says something like, "Some controlled studies have shown that things like meditation and prayer can help reduce certain conditions like high blood pressure, ulcers, and even asthma." It is valuable in these contexts to understand just what "controlled studies" means.

This is a term that has to do with running multiple sample groups. One, called the control group, does not receive the treatment or take up the particular practice being studied. Such a group provides an important contrast when the researchers run the statistical analysis. The researchers compare the treated group(s)—which is to say the non-control group(s)—and see how the variance and such are different.

To examine spiritual practices in terms of anxiety, the setup would be something like the following. The researcher and team interview all the study participants at the beginning of the program to rate their anxiety levels. This is combined with some basic health data like blood pressure, resting heart rate, and so on. Then the researcher randomly puts people into three different groups: prayer group, meditation group, and control group. The names do as they say: the prayer group prays, the meditation group meditates, and control group does neither of those things. After engaging in practices for, say, three weeks, the individuals are again asked to do an anxiety questionnaire and get new physical check-up tests done.

Now the researcher is in position to take the data that has been collected over the last few weeks and begin crunching it up and figuring out what it tells us (if anything!).

Clap your hands if you feel that happiness is for you

The following table offers us another kind of insight into religion and crosses over with some of the kinds of research that would be done in the counseling and psychology fields. Here we have data representing people's responses to survey questions on religious beliefs and reporting on their general happiness level. The bold numbers in each cell represent the percent of total respondents in each column. The regular numbers represent the total number of respondents in each cell.

Does God exist?	1 Very Happy	2 Pretty Happy	3 Not Too Happy	Row Total
1: Don't Believe	**2.5%** 147	**2.8%** 312	**3.4%** 85	**2.8%** 544
2: No Way to Find out	**3.8%** 224	**5.2%** 568	**5%** 125	**4.7%** 917
3: Some Higher Power	**9.1%** 540	**10.4%** 1,140	**9.7%** 244	**9.9%** 1,924
4: Believe Sometimes	**3%** 179	**4.5%** 494	**4.8%** 121	**4.1%** 794
5: Believe But Doubts	**13.1%** 779	**18.2%** 1,999	**16.1%** 405	**16.4%** 3,180
6: Know God Exists	**68.4%** 4,043	**58.9%** 6,457	**61.1%** 1,541	**62.1%** 12,041
Column Total	**100%** 5,909	**100%** 10,970	**100%** 2,521	**100%** 19,400

There are a variety of interesting facts about the data shown here, but I want to just focus in on one aspect in particular, and that has to do with the impact of sample size on the margin of error. The two are inversely related, which means as one goes up, the other goes down. A larger sample size means less error. Here let's consider the difference between the 147 Very Happy people who don't believe in God compared to the 4,043 Very Happy people who know God exists. This suggests that a random Very Happy person is likely to have a strong religious faith.

Let's look at how that plays out in the numbers. The formula for calculating the margin of error depends on which confidence interval you want to meet. If we aim for 99 percent confidence, then the number we plug in at the front for $Z_{\alpha/2}$ is 2.58. Again, we'd find that from a table. So:

$$E = Z_{\alpha/2} \times \sqrt{\frac{p(1-p)}{n}}$$

$$E = 2.58 \times \sqrt{\frac{0.684 \times 0.316}{5909}}$$

$$E = 2.58 \times \sqrt{\frac{0.26114}{5909}}$$

$$E = 2.58 \times \sqrt{.00004}$$

$$E = 0.0163$$

That means of all the people who would rate themselves Very Happy in the larger population, we can be 99 percent certain that the percentage who have strong belief in God would be:

$$68.4 \text{ percent} \pm 1.63 \text{ percent}$$

That's not much error, and that's because the sample size is large. Be careful, though. What this is not showing us is that if you believe in God you are much more likely to be happy. Reflect for a minute on the other values in the row for Know God Exists and think about what those other values are showing you. You should also think for a minute about the fact that Not Too Happy seems to follow a similar pattern. The largest number of people who are Not Too Happy also know that God exists. What might that indicate (if anything)? Can we determine if religion makes people happier?

Pascal's wager

We're not going to be able to use statistics to solve all the questions about religion in this book. But let's look at one of the most famous arguments about believing in God, since it is based on probabilities.

Pascal's wager is rather interesting not only for its subject, but also in that it is one of the few arguments based on probability that you will come across that uses values of infinity instead of actual numbers. The setup goes like this: there are two choices I can make and there are two possible states. This is sort of like saying you can choose door number 1 or door number 2, and behind each door is a prize. There are two possible prizes, and you're not sure which it might be.

The two choices for Pascal are: believe in God and live accordingly or don't believe in God. The two possible states are: there is a God or there is not a God. He then showed how things would turn out for us based on those two things:

	There is a God	***There is no God***
I believe	Infinite reward	Small loss
I do not believe	Infinite loss	Small gain

The results are that, of course, if I believe and there is a God, I go to eternal reward. If I believe and there is no God, then my only loss is the possible wealth and worldy stuff I could have acquired by being a Scrooge character, squeezing money from my fellow men, instead of being a good person. On the other hand, if I do not believe and there is a God, I'm pretty much toast for all of eternity. If I do not believe and there turns out to be no God, I get the small gain of Scrooge's fortune in the form of gold and fine houses and so on. (Of course, Pascal came before Charles Dickens, so he didn't have the awesome character of Scrooge to use in his version!)

So, like a good game, there are the payouts and the risks. As you might expect, Pascal concluded that the smartest move in this game is believe and live as if God exists, since the infinite reward blows away the other options. Whether you agree with his conclusion or the way he reasoned through the issue (there are many brilliant thinkers on both sides of this debate), I hope you can appreciate what a delightful and clever piece of mathematical thinking it represents for tackling a very tricky problem.

Crime

If religion is the highest we can reach, then crime is, by contrast, the lowest we can fall. Statistics and probability are not useful instruments if they cannot navigate in the dark areas of the forest or the deep areas of the ocean, where fish look like there are just giant sets of teeth with a little tail pinned on.

Knowing the victims of homicide

One of the most forceful ways that our perception of crime and safety in our communities can be affected is to personally know someone who has been a victim of a violent crime. Something like a simple survey and the makeup of a sample group can strongly impact means and medians. These variations can help us understand why people attempting to engage in dialogue about political and social issues can be talking about two different worlds.

Consider the following example: In a survey conducted within the last five years that included a large metropolitan area—the inner city, suburbs, and some of the small towns in the "antique store radius," which is where people will drive out to spend a day hitting small, old shops and imagining a quieter, quainter version of America—the results showed:

Number of People You Have Known Who Were Victims of Violent Crime

# of Victims	*# of Respondents*
0	3,943
1	278
2	96
3	41
4 or more	24
Total	*4,382*

The mean here is going to be:

$$\text{mean} = \frac{(3943 \times 0) + (278 \times 1) + (96 \times 2) + (41 \times 3) + (24 \times 4)}{4382}$$

$$= \frac{0 + 278 + 192 + 123 + 96}{4382} = \frac{689}{4382} = 0.157 \text{ victims}$$

That's kind of a weird statistic, but it describes the average. Notice how that's using kind of a "best case" scenario by setting the number for "4 or more" at its lowest possible value of 4.

Another common statistic for representing the middle of a group is the median. The general form of the equation to find the median for an even number of things is:

$$\text{median} = \frac{\frac{n}{2} + \frac{(n + 2)}{2}}{2}$$

Here *n* represents the total number of values in the group. For example, *n* could count how many kids took a test or how many people answered a survey. The median allows us to find the middle position if we lined up all the values from smallest to largest. When we plug in the numbers from the crime survey, we find:

$$\text{median} = \frac{\frac{4382}{2} + \frac{(4382 + 2)}{2}}{2} = \frac{2192 + 2192}{2}$$

$$= \frac{4383}{2} = 2191.5$$

Since position 2,192 is within the first group of scores (3,943), the median is 0. Do you think that is going to be a meaningful statistic?

What happens when we compare the two methods of measuring the middle—mean and median? The mean of 0.157 victims would suggest that, in theory, if you got a group of seven citizens together, someone in the group would have known a **homicide** victim. In contrast, imagine if you are running for re-election and you want it to seem like you have made the city really safe. In that case, you would probably point out that 89.9 percent of your citizens have not known even a single victim of a violent crime over the past year. The median, after all, is zero—a vast majority of people know no victims.

However, and it's a big "however," notice what happens if your political opponent takes the same survey data but shrinks the radius to only include the city proper and the suburbs that directly touch on it. Chances are good, we can agree, that the overwhelming majority of the respondents no longer being counted are from that peace-loving, quiet outer layer. If we have only 3,000 people in the first category of knowing no crime victims, then the new mean becomes:

$$\text{mean} = \frac{(3000 \times 0) + (278 \times 1) + (96 \times 2) + (41 \times 3) + (24 \times 4)}{3439}$$

$$= \frac{0 + 278 + 192 + 123 + 96}{3439} = \frac{689}{3439} = 0.2 \text{ victims}$$

It went up from 0.157 victims to 0.2—a jump of 27 percent! More importantly, your opponent can then quote values like 12.76 percent of the citizens have known one or more homicide victims. Even worse, he might use the same strategy as before and claim that the 0.2 means if we took a random group of five people from our city, one of them probably knew someone who was a homicide victim.

Does the smaller radius make the median score meaningful? The formula is slightly different since it's an odd number.

$$\text{median} = \frac{n + 1}{2} = \frac{3439 + 1}{2} = \frac{3440}{2} = 1720$$

The value in position 1720 is still going to be 0. It's a noticeable difference in position, but we're still at a value of 0. This is a good example where the median is not a meaningful (or even useful) measurement.

Keep in mind we're still working with a kind of imaginary value, because no one—at the same time—lives in a high-crime neighborhood on the south side of the city *and* a tree-lined suburb. If we run the numbers again, cutting out the lowest-crime suburbs who have their own separate police forces, we would start to see a much different set of numbers. We won't walk through the formulas over and over again, but just reflect on how different the world looks if you are from a quiet suburb versus a neighborhood where most of the people you know, know someone who was the victim of a violent crime.

Jury duty

One of the ways that many people come into the most direct contact with the criminal justice system is through the process of being randomly selected for jury duty. The use of juries is a key ingredient in helping to make sure that people receive a fair and objective trial, so it is probably not surprising that a lot of effort is taken at different stages to select a fair and objective group of people to serve on a given jury. Statistics can help with this, because there are many times that it is hard to tell whether something that seems fair or unfair actually *is* fair or not.

Let's consider a situation where a jury ends up with no women on it, even though the city in which the trial and jury selection are taking place has a balanced population with 50 percent men and 50 percent women in the eligibility pool. In order to test out the defense attorney's complaint that the jury was tilted toward men, let's look at the probabilities.

First, we could check to see what the numbers would look like if the process were truly random. The number of outcomes works

The elevator paradox

It has been observed over and over that if you are waiting on a floor near the top of the building, an overwhelming majority of the time the first elevator to stop at your floor seems to come up from below. However, if you are waiting at a floor near the ground, the reverse is true: an overwhelming majority of the time the first elevator to stop at your floor appears to have come from above. If most of the elevators to the top come from below and most of the elevators near the bottom come from above, that would give the impression that there are elevators being created in the middle of the building.

There is an easy way to get our heads around this, though. Imagine a fifteen-floor building. Let's allow that the probability of stopping on any given floor is equally likely. If you are on floor two, there is one chance in fifteen floors the elevator will be at your level when you buzz for it, and there is a 1/15 chance it can come from below you. However, there is a 13/15 probability the elevator car last traveled to some floor above you. So, it is possible that the elevator could come from first floor to the second; however, it's unlikely.

By evaluating things like who's on a jury or which people are being arrested, statistics help us make sure laws are carried out fairly.

like flipping a coin. With two possible outcomes each time, we have $2 \times 2 \times 2 \times \cdots \times 2$, or 2^{12}, which equals 4,096 possible outcomes. Keep in mind MMMMMMMMMMFM is a different outcome, in this model, than MFMMMMMMMMMM. This happens similarly with all the possible sequences that have two women or three or eight.

There is only one possible outcome where the jury has no women, so the probability is 1/4096, which equals 0.024 percent.

That is *extremely* unlikely. Of course, that doesn't necessarily prove anything. Extremely unlikely things do happen in the world. People get hit by lightning. People win the lottery. Sometimes you can even flip a coin and get twelve heads in a row. Fortunately, in cases like jury selection, there are some mechanisms available to allow us to adjust the mix and force a slightly more representative group to be seated for the trial.

Lurking variables

In working to analyze the variables that match up with high crime rates, one of the things that seems to play a part is education, as shown in the following graph.

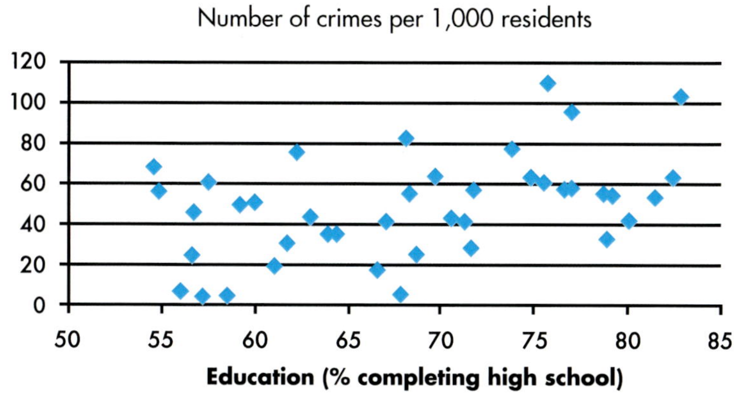

Correlation measures how closely two variables are related. Correlation is measured with the variable *r* and ranges from –1 to 1. If the correlation between two variables is 1, that means as one goes up, the other goes up proportionally. For example, if you buy gas for a car, the total price of the gas is perfectly correlated with the number of gallons you buy. A correlation of –1 means as one variable goes up, the other one goes perfectly down. If you car gets a constant 25 miles per gallon of gas, for every 25 miles you drive, your total gallons will go down by 1.

In the above crime graph, we see a positive correlation between the number of people in a given municipality who finished high school and the crime rate. That means if a community has more people who finished high school, it is likely to have a higher crime rate. That's interesting because that tends to go against our expectations. In fact, the correlation between the two variables is $r = 0.437$. Should we trust the numbers and accept the conclusion that more education causes more crime? We've certainly seen other cases where the numbers helped to fix our misperceptions; maybe this is one of those times.

Well, it might be, but when we have trouble coming up with a **plausible** reason that might account for it, we should be willing to dig a little deeper before going along. One way to do that is to look for other variables that might also have a very strong correlation to both of these. Sometimes the similarity is because of a common variable that is hiding in the background. We call this "lurking." (How can you not love it when someone picks a good word like "lurking" for a technical term, instead of something boring like "under-indicated influence factor?")

When we do that, we find that there is an even stronger correlation between crime and urbanization (how city-fied each of the areas is), as shown in the graph below. The urbanization numbers on the x-axis represent the percentage of people in an area who live in a city.

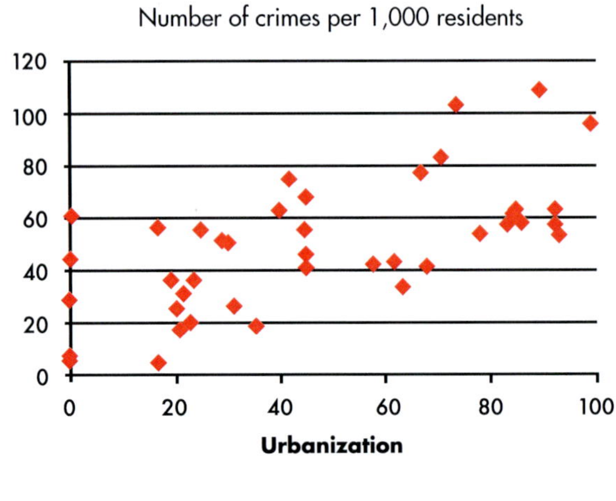

Not only do we see a similar correlation here, but it is stronger, with $r = 0.662$. There is also a positive correlation between the other two variables—urbanization and education—but looking at that one wouldn't add much to the discussion.

This puts us in a position to be able to explain why it seemed like education resulted in an increase in crime and gives us an example of an important rule in statistics. We can understand how high school completion rates tend to be higher in city areas than in rural one. We can also understand some of the reasons crime tends to be higher in city areas. The famous saying in statistics is, "Correlation does not equal causation." That's the technical way to say it. In normal English, we can say, "Just because two things happen together doesn't mean one of them is causing the other one."

Really bad property damage from fire tends to match up with having a large number of firefighters at the scene, but I think you can see that the property damage is not *caused* by having fire trucks and firefighters present. There is a lurking variable there, too. Another example: if I eat a bunch of cookies for dessert one night, I might want to burn off those extra calories by running faster during my workout the next day. But does that mean cookies actually make me faster?

One missing city

If we look at the table on page 54, it shows the data analyzing the rate of violent crimes per one hundred thousand people for a given year. In group one, all the cities are included, but in group two, one of the cities has been left out. Based on the differences in the numbers, what conclusions would you draw about the level of crime in the city that was taken out? Did it have a higher or lower crime rate than the average? If it did, was it a lot higher/lower, or just a little? Why would the means change even though the median value stayed the same?

	51 cities	***50 cities***
Mean	5.6	4.8
Standard deviation	6.05	2.57
Maximum	44	13
Median	5	5
Minimum	1	1

Politics

This category will show up in the next chapter as well. It's valuable to try to break it in two to show how it is different from the side of voters versus politicians.

Positions along the spectrum

It often feels like the changing political climate of our country and community can be as **tangible** as changes in the weather. It can impact our lives at least as visibly (and often in more lasting ways). Because of this, one of the most common and most watched kinds of surveys of public opinion is the attempt to gauge where citizens' political mood stands.

In a survey allowing people to pick from a numbered range from 1 to 7, where 1 is "extremely liberal" and 7 is "extremely conservative," a survey came up with the following set of responses:

Score	Category	# of Respondents
1	Extremely liberal	69
2	Liberal	240
3	Slightly liberal	221
4	Moderate	740
5	Slightly conservative	268
6	Conservative	327
7	Extremely conservative	68
	Total	***1,933***

This kind of survey uses what is known as an ordinal scale. It is not really the same as a **quantitative** measure, but it can be treated and crunched in ways as if it were by assigning scores to each of the categories and then finding things like the mean, mode, and median.

In this case the mean is 4.11. What would that tell us about the general mood of the community surveyed? How would the "average" citizen describe herself politically? The mode here is moderate (4) and seems moderately (ha!) useful because the grouping is rather symmetrical with no strange lumps on either side of the mean.

The fault is in our votes

Here it will be interesting to look at how statistics can be used to help us understand when things seem to go wrong with the voting process (and how to check to make sure it hasn't gone wrong). What we see here is an example where there appears to have been

Meet Linda Qiu

I was fortunate to catch up with Linda Qiu, one of the amazing journalists at politifact.com. Linda has a college degree from the prestigious University of Chicago and experience writing for *National Geographic*, among others.

Linda, politicians use statistics pretty often, so I'm curious to hear how much you think they are being used well or used badly.

For the most part I don't think politicians are trying to mislead anyone intentionally. It is common, though, because they are so passionate about the issues. They will use the statistics in a way that ends up being misleading.

A good example is using a statistic about, say, how world poverty has been affected by a certain group or how unemployment has gone up or down in the past few years. In those cases, while the statistic might be technically true, what is being counted in their definition of "poverty" or "unemployment" has shifted, so it is no longer actually comparing the same things.

Another great example is when a candidate latches onto survey results that really help their case, but didn't dig down to see if the organization doing the survey followed best practices in the first place.

Is there something you particularly like about working with numbers so much more than a regular journalist might?

Oh, sure! The statistics allow us to cut through a lot of surface noise and get to the heart of an issue. Sometimes we look at a claim like, "I am getting calls from voters whose health care costs have risen 30 percent or 50 percent." In the context of a bigger picture, though, it turns out that those are **outliers** (or nearly so), and the average for the group is closer to something like 7 percent or 10 percent.

What would be your dream statistic? What do you think would help you and other voters evaluate candidates more accurately?

Honestly I'd have to put in a plug for Politifact. It is heartening to see how our work, and other news organizations that have now launched similar programs, are becoming a part of the public discussions to help voters get a more accurate version of things being claimed. Politicians will even mention us (when they are right or when the opponent is "pants on fire").

Thank you for sharing your thoughts with us, Linda. Best of luck in your work trying to keep politicians honest and voters well informed!

some irregularity with the way ballots were marked, and how much of an impact it would have on the results of the election.

The below graph shows how many votes Candidate A got in the last election in each county compared to how many votes Candidate B got in this election. The graph shows that there is a strong correlation between votes for Candidate B this time and votes for Candidate A last time—the candidates held similar positions and belonged to the same political party.

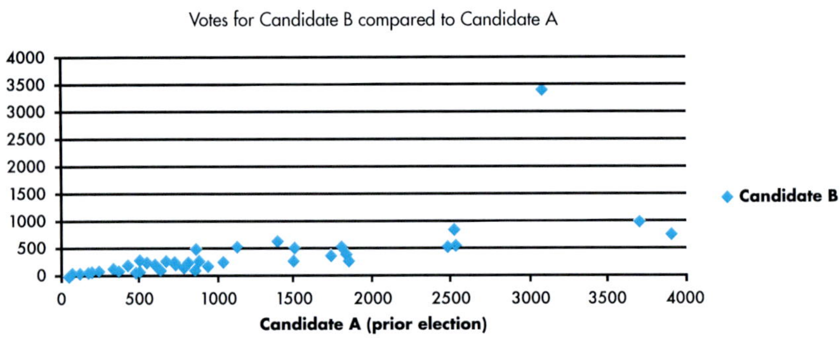

In one county, there is a wild leap between voter preferences this time versus last time, and there is no obvious explanation (such as if the candidate were from that county, causing voters to vote for him when they otherwise might not). This data point, a large variation outside of what could reasonably be expected, is known as an outlier.

If we compare the correlation values with the outlier left in, we get a value of $r = 0.713$. In contrast, if we calculate the correlation with the outlier taken out, the value goes up to $r = 0.921$. That means a jump from 71 percent to 92 percent. This strongly suggests that something went wrong with the voting in this particular county.

In the event that the political contest was decided by less than the margin of what seem to be voting errors, there would be grounds for a drawn-out court battle to resolve the issue and try to make sure the irregularity was corrected.

Cars

Here is my one concession to an actual vehicle we use for traveling around seeing and exploring stuff. It is a peculiar feature of modern society how much our cars are a kind of extension of ourselves—our sense of freedom, our style, our social status. If there is one possession that **encapsulates** more dimensions of our psychology, and our perceptions about other people, than our cars, I can't imagine what that would be.

I know some of you reading along are starting to feel all smug and are thinking, "No, cars aren't an extension of *my* self-esteem." Well, I hate to break it to you, but there is something inescapable here, because your attitude about cars still reflects a whole bunch of stuff about your self-image and your image of other people.

Big cars, big gas problem

Does the weight of a car or truck impact its fuel efficiency? If we begin by listing out a set of data, we might get results from a small sample group that looks like this:

Car	Weight (lbs.)	Miles per gallon
Sedan Z	3,164	34
Family Wagon	2,590	38
Big Black Truck	3,838	22
Yeti 4WD	3,970	21
Huge Truck	6,400	17

We'll plug that data into the graph on the next page.

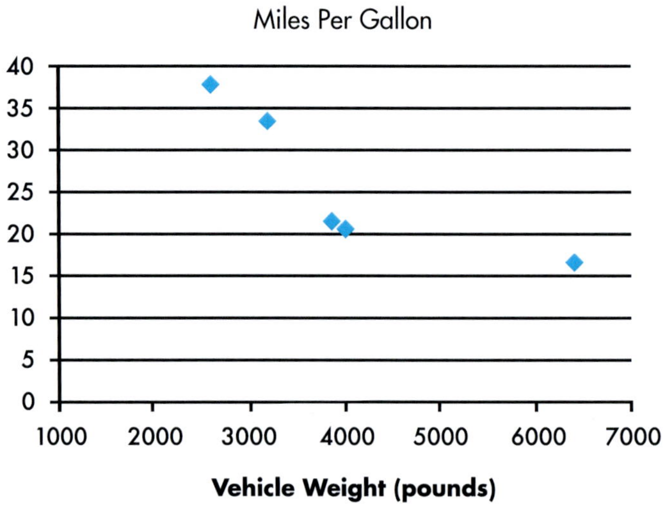

Let's say we wanted to build a forecast model where, given a car's weight, we could predict its fuel efficiency in miles per gallon. We could try to build a linear model based on something called "regression." It is called that because the model draws a line where all data points regress to, or are pulled toward, the mean. Before telling you what the line of regression is based on calculation, take a minute to think about what you would put as an estimate. The form would be:

$$y = mx + b$$

where:
y = The vertical axis value (miles per gallon).
m = The slope of the line that is drawn through the midst of the dots.
x = The weight variable along the x-axis.
b = The y-intercept (where the line would cross the y-axis if $x = 0$).

Remember, the formula for the slope of a line is:

$$\frac{y_2 - y_1}{x_2 - x_1}$$

Reasonable end points to use for the line might include (2590, 38). The second one would be estimated between the last two data points. At least think about it a little. Would the line have a positive or negative slope? Would the slope be a large or small number? What would be your estimate for where the line would hit the y-axis (when $x = 0$)?

OK, if you've actually thought about it, I'll give you that the formula is $y = -0.0052x + 47.32$. This means, generally speaking, that the heavier the vehicle is, the less fuel efficiency it has.

You shall not pass

If three fifteen-year-old friends decide to go on the same day to take the written portion for their driver's exams so they can get learner's permits, and each exam is either pass or fail, how many possible outcomes are there?

This works like flipping a coin three times, since there are two possible outcomes for each friend. The number of results that could happen is $2^3 = 8$. Since only one of those outcomes results in all three passing, the probability is:

$$P(x) = 1/8 = 0.125$$

Suppose that in this state about 70 percent of future drivers (not 50 percent) pass the written test on the first try. If these three are a good representative sample, what would be the probability that all three pass, then? Since they are independent events, each person has a 7/10 chance of passing:

$$P(x) = 0.70 \times 0.70 \times 0.70 = 0.343$$

I'm too tired

In thinking about what is valuable and meaningful in statistical comparisons, consider a company running a test on two new tire prototypes to see which one is better. In both cases, the measurement is in thousands and shows the number of miles before the tread has worn down to the safety limit.

Prototype A	Prototype B
48	46
49	53
51	54
52	47

Both sets have the same mean and same median. Ask yourself why prototype A might still be considered a better, safer tire. (Hint: Think about another common measure we've calculated with other data sets. It's pretty standard.)

Environmental Activism

While we're wandering around inside a big metaphor about exploring the world around us with all its scary-cool forests and deep coral-reefed oceans, let's not forget to include the process of protecting the actual planet. We still want to have amazing places to explore when we aren't sitting around reading about how awesome math is. Part of that includes protecting the creatures that run and swim and fly all over the world, helping to create the living, changing ecosystems.

Manatees are gentle, curious creatures. They are an endangered species, and are often killed due to collisions with boats.

Manatees

In Florida, the large, slow-moving manatees are endangered by increasing levels of boat traffic each year. We could collect data every year comparing the number of boats registered and the number of manatees that die from encounters with boats. If we translate that to a scatter plot graph, we are able to make a meaningful guess about the existence of a correlation as well as an estimation of whether the correlation is strong or weak.

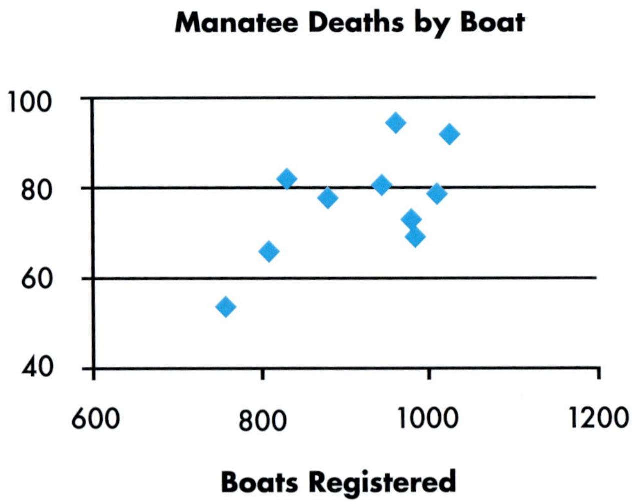

Looking above, we see a positive correlation and likely a fairly strong one, judging from the how closely placed the plots would be along the imaginary line we might sketch through the middle of them. The line would have a strong upward slope.

How might we make some decisions based off of this to help reduce the number of manatee deaths next year? Naturally, limiting the number of boats that can be registered might achieve part of the goal, but what are some other possible options? Are there possible regulations that could make the boats less of a risk to the manatees, perhaps?

Airborne toxic event

One of the ongoing battles for the environment on the grandest international scale is the struggle over carbon dioxide emissions. On the next page is a standard box plot of the emissions in metric tons for various European Union (EU) countries. It shows the mean (the line midway through the box), the upper and lower values (the crossbars at the ends) and the seventy-fifth and twenty-fifth quartiles (the sides of the solid blue box).

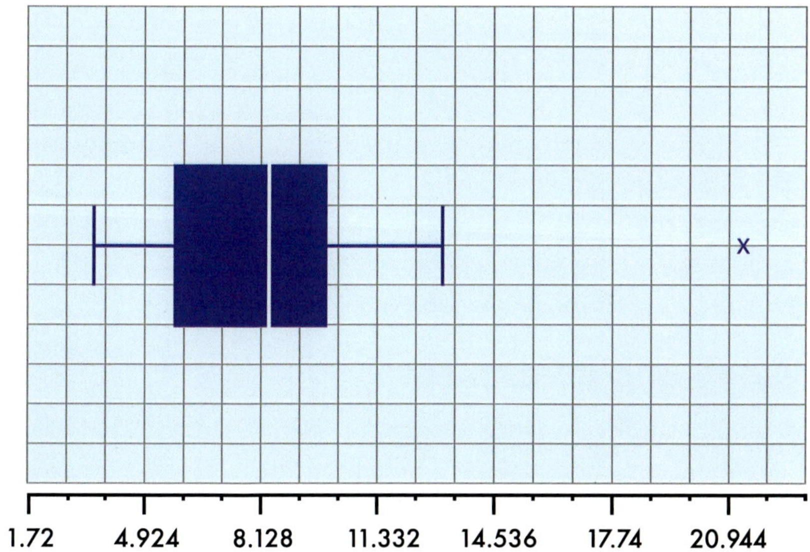

If the mean is 8.3 with a standard deviation of 3.6, then we can ask a few questions to understand what we're seeing:

1. How many standard deviations from the mean is the outlying value of 21.3?
2. If three standard deviations is the limit beyond which a data point is considered an outlier, is the United States, with a value of 18.9 for that year, an outlier?

Well, in answer to question one, we would calculate the following:

$$\text{Number of standard deviations from mean} = \frac{\text{outlier} - \text{mean}}{\text{st. deviation}} = \frac{21.3 - 8.3}{3.6} = 3.6$$

So the maximum value is an outlier, but what about America?

$$z = \frac{18.9 - 8.3}{3.6} = 2.9$$

Well, since the normal dividing line for what counts as an outlier is three standard deviations, it seems the United States would not technically be an outlier, but would probably fall into a gray area that merits some additional focus without falling outside the main group in most considerations of policy.

Wealth and Investment

We tend to think of wealth as a kind of "thing" that we can possess, or as the means of possessing a bunch of things. And it is, at least for some people who have lost their perspective and their sense of real adventure. Instead, perhaps let us think of it as something like the liquid that fuels our submarine or airplane or spaceship, and allows us to take our mind on adventures. I mean, just think: you can even use your wealth to buy books about imaginary places or, you know … math! It's also fascinating because, similar to cars, people's attitudes about money and their choices of what to do with it can tell us worlds of stuff about them (and us).

Is the class half-full or half-empty?

An interesting bit of data comes out of research on the effect of wealth and whether that is passed on to children—even though it's not actually genetic. To illustrate this, let's consider the following table.

The table on the next page reads left to right, and gives data about the economic class children grow into as adults depending on their parents' economic class. For example, if you were born into the middle class (the middle 60 percent of all incomes), you would have a 15 percent chance to move into the upper class, and a 67 percent chance to stay in the middle class. But rather than analyze the different numbers, I want to offer the following

	Economic Class of Children		
Class Level of Parents	Top 20%	Middle 60%	Bottom 20%
Top 20%	43%	51%	6%
Middle 60%	15%	67%	18%
Bottom 20%	8%	57%	35%

graph so you can get an intuition for what these numbers show us. Sometimes in math, contrary to how we are usually taught, there is an insight or an intuition that strikes us, and then we have to figure out how to explain it mathematically. I will leave it to you to reflect on how you would break down what is meaningful in the graph:

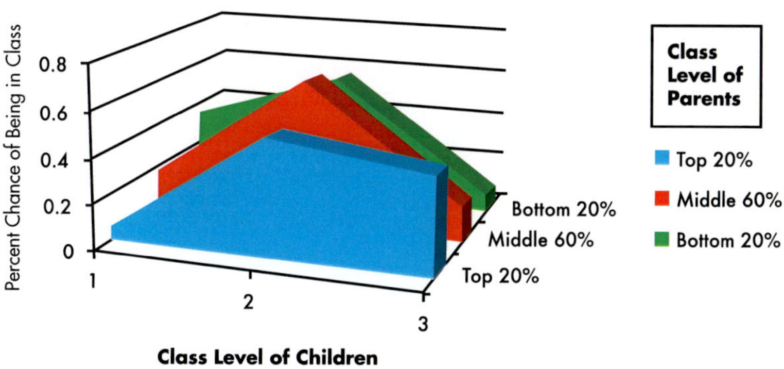

A 1 in this chart means the children who end up in the lowest 20 percent, 2 means the middle 60 percent, and 3 is the children who end up in the top 20 percent. Can you see how the lump moves as we switch from green to red to blue like a wave moving from left to right? What numbers or values do you think would help us interpret how much movement has occurred?

Does money buy happiness?

Here is an example of using a simple chart of proportions to look for any sort of trend or pattern that might merit an expanded research project.

	Life Satisfaction			
Income	***Not Too Happy***	***Pretty Happy***	***Very Happy***	***n***
Above Average	$\frac{26}{423}=0.061$	$\frac{233}{423}=0.551$	$\frac{164}{423}=0.388$	423
Average	$\frac{177}{883}=0.133$	$\frac{473}{883}=0.536$	$\frac{293}{883}=0.332$	883
Below Average	$\frac{182}{687}=0.265$	$\frac{357}{687}=0.520$	$\frac{148}{687}=0.215$	687

In looking at the change in values, we can see here that likelihood of "very happy" appears to climb as we move up the income scale. The same seems to be true for "pretty happy" as well. But the trend is reversed for the "not too happy" group. That measure goes down as you move up the income scale. Think through this a little bit, though. What is really being measured in this column? Would it make sense to trend upward with income or downward?

What is the risk of taking a risk?

There is a whole interesting field that has grown up at the intersection of psychology and economics, called behavioral economics, and the following example will give us a nice chance to consider how we add some complexity to a situation that is not actually there in the numbers themselves.

Consider a situation where you receive \$1,000 to invest. However, you have to choose between two options. In option A, you get \$500 as a sure thing for the investment. For option B, you have a 50 percent chance of getting another \$1,000 and 50 percent chance of getting \$0 additional dollars. When faced with two possibilities, we can calculate their expected value by multiplying each payoff by its probability and adding the results. That means the expected gain in option B is:

$$(\$0 \times 0.5) + (\$1{,}000 \times 0.5) = \$500$$

Option A and option B both have an expected value of \$500. But in studies, when this kind of scenario is run again and again, people tend to pick option A, the sure bet. They are averse to the possibility of getting \$0 from the project.

In a second situation, however, you receive \$2,000. This time you must choose between two different losing options: In option A you lose \$500 for sure, but in option B you have a 50 percent chance of losing \$1,000 and a 50 percent chance of losing \$0. Again, the expected loss works out the same for both options:

$$(-\$1{,}000 \times 0.5) + (-\$0 \times 0.5) = -\$500$$

Interestingly, when faced with this sort of scenario instead, people predominantly pick the risk-taking strategy (option B) instead of the sure-fire strategy.

Even when none of the money was ours to start with, we behave and decide very differently when it seems like we are maybe gaining versus situations where we are maybe losing something. How do you think you would choose from the options offered?

Happy as Scrooge

There have been a variety of clever experiments showing that we derive more pleasure when we spend our money on shared experiences with people than when we spend it on things. This happiness is both at a higher level and longer lasting. So, for instance, taking a vacation with someone tends to have a bigger impact than, say, buying a new car. (Yes, even though I talked about how cars are a big deal. Remember, I said they were a big deal on the list of *things*.) Similarly, your friends and your kids will hold fond memories of an afternoon baseball game long after they have forgotten what toy you bought them for their eighth birthday.

Even more amazing, though, are the wonderful bits of research that have been able to show that we derive the greatest amount of happiness by giving to others. In addition to testing this theory in a variety of different countries and with a wide range of possible things to give away, whether actual money, bags of chocolate, or gift cards, the results seem to be pretty much universal. In my favorite example, researchers set it up to see if it was true even for children just under the age of two. That's old enough for the experiment to work, but young enough to rule out a lot of cultural conditioning. This would help them to see if it was really a feature of our biological hardwiring, if possible.

With trained observers watching to score the kids' reactions, the experimenter went through a series of steps. After introducing the children to a little puppet, the experimenter gave the children a small bag with crackers. Then they asked the children if they would give a cracker to the puppet (who ate the treat with lots of happy sounds and actions). Then the experimenter "found" an

Actuaries

Actuarial science is a specialized field within mathematics that is focused around the assessment and minimization of risks. In particular, actuaries often work with insurance companies. There they help analyze the different variables having to do with life expectancies, health risks, and medical expenses. There are also a number of actuaries who will work for financially centered businesses.

Actuaries work with computer software in order to pull together the chances that an event will occur, to determine the probable financial cost to the insurance (or other) company when it does happen, and to assess the appropriate cash reserves to keep on hand.

All the different aspects of this come together with some **assumptions** about the business, customer load, and more, in order to compile the complicated pricing structure and coverage amounts that show up in insurance policies.

The roots of actuarial science stretch back into antiquity, with different elements of the current model coming into play at different points. There were precursors to insurance companies in the form of mutual aid societies as far back as the fourth century BCE. But it was not until the 1660s that a draper named John Graunt used math to actually compile the first form to apply for insurance.

extra cracker and handed that to the child, asking them to give it to the puppet.

Here is a chart showing how the events were rated by the trained observers:

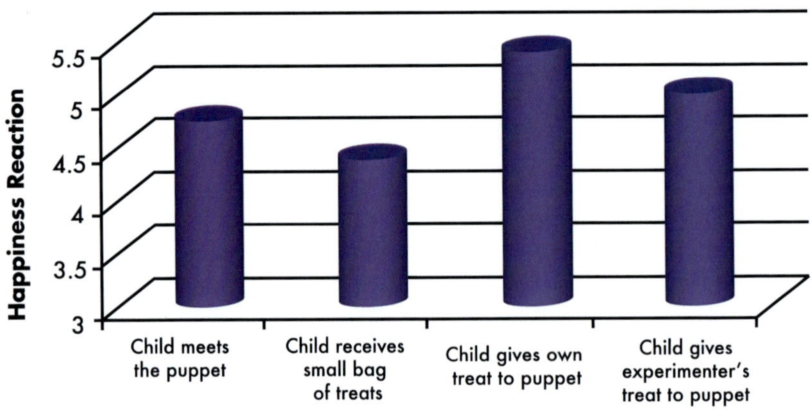

Notice that the children were even happier when they gave away one of the crackers from their own stash than they were when they gave away the "extra" one the experimenter found. And so we see how Scrooge can wake up a changed man who starts giving out figs.

Across the Universe

As we round out this chapter, let's finish by looking out toward space, considering how statistics can help us make sense of some of the strange uncertainties that confront us in the infancy of our exploration out across the universe.

The Drake equation

The first of the probability puzzles I want to share with you is called the Drake equation. The challenge at hand is the attempt to formulate a good theory or explanation concerning the chances that there is intelligent alien life out there somewhere. If your curiosity is sparked and you go out to do more research on this topic, you will find that there are several different versions of this equation. The approach I tend to favor tries to hit a middle ground between versions that are overly conservative with their estimates and those that are a little too liberal in allowing for high likelihoods of everything.

The idea itself is not particularly complicated, and it works much like calculating the probability of rolling a one on a die three times in a row, or of drawing a heart and then a spade from a deck of cards. All such cases work by multiplying the probabilities of each event. In this case we have a long equation of probabilities that need to get multiplied in:

$$N = R f_p n_e f_e f_i f_c L$$

where:

N = The number of intelligent civilizations in our galaxy.
R = The average number of new stars that appear in our galaxy per year = about 7.
f_p = The fraction of those stars (R) that have planetary systems = about 0.4.
n_e = The number of f_p's that could support life = around 1.
f_e = The fraction of n_e's that actually do develop life = 0.5.
f_i = The fraction of f_e's that develop intelligent life = around 0.5.
f_c = The fraction of f_i's that go on to become highly civilized = 0.5.
L = The number of years f_c's would likely exist before destroying themselves or being destroyed from outside = 10,000.

$$N = 7 \times 0.4 \times 1 \times 0.5 \times 0.5 \times 0.5 \times 10{,}000$$

$$N = 3{,}500 \text{ estimated intelligent civilizations}$$

The Parkes Radio Telescope in Australia has been collecting information about our universe for over fifty years.

That's a little mind-boggling every time I stop to think about it. Keep in mind, this is just calculating the probability for civilizations that are currently existing at that intelligence level. There could have been tons more since the earliest planets formed in the universe, and there could be tons more that are out there now but have not yet reached the stage of being highly civilized.

The probability of the universe existing at all

There's one last idea to touch on. What is the probability that our universe came to exist in such a finely tuned way to be able to support intelligent life at all? If any of the features of our universe had developed in just the tiniest way differently than it did, we would not exist. But a kajillion other things would or could be different in other possible versions of this universe. So isn't it amazing that the odds worked out just exactly right for our universe to exist?

Well, the interesting problem with that theory has two parts to it. First, the probability of anything that has happened is always 1 (that is, 100 percent likely to have happened). Second, in order to calculate a probability, we have to compare it to other universes. But we can't; we only have one data point. Therefore, we can't determine the probability of our universe existing. In spite of our various scientific theories, we have not yet been able to test out and verify the existence of other universes, nor do we have a clear sense of how many universes there might be in total—how many possible outcomes there are. If we don't know how many sides there are on the dice or how many tickets are in the lottery, then there's no way to set the odds.

So, there we have managed a sampling tour of some of the different vantage points from the circuit of our daily lives, and we have tinkered with a variety of different possible tools and instruments that we can use to get a reading on ambiguous things we experience. In the next chapter, we'll look at how others use these same tools in specialized ways.

Statistics can help teachers understand how well the teaching process is going and make sure they are measuring students' work accurately.

THREE

Statistics and Probability in Others' Everyday Lives

In this chapter, let's see if we can imagine what it's like to look out at the world from the vantage point of various other people. Instead of seeing the world through the filter of just our experience, now we will also try to get a sense of how it looks through someone else's filters. The beauty of this is that not only does it give us some additional exposure to different ways math can be applied, but it also broadens your mind by seeing examples of processes and puzzles that probably don't come up in your normal daily experience.

We will start by flipping over to the other side of the desk, so to speak, from where we began in chapter two. We will do something similar by switching into the shoes of a law enforcement trainer, and a politician as well. Along the way, we will imagine ourselves to be people who try to use statistics to outguess the chaos of investments or the shifting weather and climate movements.

Remember that to get the most out of it, you have to go in with a sense of curiosity and an open mind. Just allow yourself to think, "I wonder if I'll find anything cool or strange in here." Those curiosities are what spark us forward, and even just pretending to be curious can open up the creative problem-solving parts of the brain!

Teachers and Professors

Teachers and professors working at different levels of the educational system share many of the same features of their jobs which center around primary responsibility for the design, execution, and assessment of instructional material. The breadth of curriculum areas will differ depending on the grade level, as will the materials and modes of teaching. Teachers' and professors' roles include not only the organization and presentation of material, but the design and grading of various assignments and tests to aid in the mastery of course content.

Beyond the narrowly defined instructional time, teachers and professors will also likely have numerous responsibilities and roles with respect to student organizations, administrative functions of the school, and interaction with parents and other stakeholders regarding each student's progress and any obstacles interfering with her success.

Online workshop work

A very common role for statistical analysis in education is an evaluation between two different tests to see if some process or method helped to create a significant improvement in student performance. This is called a two-sample inference.

The two groups of scores below represent students' performances on math quizzes before and after they participated in an interactive online workshop. They are from the same students, so other variables are controlled for.

Student	Ashley	Amy	Brady	Lizzie	Jonah	Sean	Marissa
After	66	68	74	88	89	91	100
Before	50	55	60	70	75	80	88
Improvement	16	13	14	18	14	11	12

Let's consider how each student improved by using the online workshop. The sample mean for the improvement scores is:

$$\frac{16+13+14+18+14+11+12}{7} = 14\text{, which we'll label as } a \text{ for average.}$$

To find the standard deviation (S_d) of the improvement scores, we calculate:

$$S_d = \sqrt{\frac{\Sigma(x - a)^2}{n - 1}}$$

That just means we will add up all the squares of the improvement scores minus their average, like this:

$$S_d = \sqrt{\frac{(16 - 14)^2 + (13 - 14)^2 + \cdots + (12 - 14)^2}{7 - 1}}$$

$$S_d \approx 2.3805$$

If we want to construct a 95 percent confidence interval (the range within which we're 95 percent sure the mean of *all* students improvement scores would fall) based on this sample set, we'd look up a factor in a table and see that $t_{\alpha/2}$ is 2.447. In order to calculate the interval, we'd plug our information into a confidence interval formula used for small sets of data:

$$CI = a \pm t_{\alpha/2} \frac{S_d}{\sqrt{n}}$$

Doing the math, we get 11.7983 and 16.2017. That means we have 95 percent confidence that that average change in scores for all students who use the interactive online workshop would fall between about 11.8 and 16.2. This shows that we would expect the workshop to increase all students' scores by at least 11 points!

No chance, Mr. Chancellor

Sometimes state politicians will attempt to influence the state's university systems. For example, a chancellor, who is the head of the university system, might say that the correlation between good teaching and good research seems to be near zero.

The problem with this line of thinking is that the chancellor seems to have misunderstood two pieces of the puzzle. The first is that there is a lurking variable here, which is the time available. Professors might not have enough time to achieve both excellent teaching and excellent research. The second problem is that there is likely a negative correlation between quality of research and teaching, for as the time is spent to strengthen the research dimension of their work, professors often have less time to devote to teaching, so the scale of their teaching impact goes down. We can see from that fact that the two variables are connected—increasing one decreases the other. The chancellor thinks they are disconnected.

A triple play of test groups

What happens if we want to compare performance between three different class formats? Let's say the college was testing out some options to begin building their online offerings, so they want to test between a traditional class, a class that is entirely online, and a **hybrid** course that has some in-class portions and some online segments (maybe they meet in person once each week, and work online in between).

Let's look at a random sample of six students from each class section to see how the grades turned out when they took the same exam at the end of the course.

Traditional: 75, 75, 95, 60, 60, 80, with a class mean of 74.1667 and a standard deviation of 13.1972.

Online: 70, 75, 60, 90, 85, 50, with a class mean of 71.667 and a standard deviation of 15.055.

Hybrid: 95, 60, 90, 75, 85, 75, with a class mean of 80 and a standard deviation of 12.6491.

We can run statistical tests to determine whether or not one class format was more or less successful than the others. When doing these kinds of tests, we begin with what is known as a "null **hypothesis**." We assume there are no differences between these sets of data until we can statistically prove that there are some. As it turns out in this case, there is not enough evidence for us to be able to say there was a substantial difference in the means between the three groups. That means we stand by the hypothesis that there was no meaningful difference. In the case of building out the program, then, that is good, because it means we can show that students get the same benefits without having to come into campus three times a week and sit around in the classroom together.

One thing to mention is that when you run a three-way analysis like this, each of the groups has to be measured against each of the other groups to be sure that they are not significantly different. We'll see a little more detail about this calculation process when we talk about water reservoir levels in the later section about weather.

Grade distributions

Using a method that is very similar to the one we saw for figuring out what a student needed to get on the final exam, teachers are constantly using statistics to evaluate and adjust the distribution of grades for tests or projects as well as for the course as a whole. This allows them to make sure that they are not expecting too much or too little compared to what students are capable of achieving.

Let's look at an example where the teacher wants his class to fall into the following groups:

A = 20% = The top 20% of the class.

B = 40%

C = 30%

D = 10%

The grade point values are 4, 3, 2, and 1, respectively. In order to see where the average grade for this kind of distribution would fall, we need to map these percentages to the grade point values:

A = 20% × 4 = 0.8

B = 40% × 3 = 1.2

$C = 30\% \times 2 = 0.6$

$D = 10\% \times 1 = 0.1$

Mean = 2.70, which puts it right around C+.

This gives the teacher a meaningful value that he can use to make adjustments, if needed. The teacher could compare his students' semester scores with this average. He could see how much or how little (and in which direction) to change the lesson plan and the test difficulty to help hit the target zone for challenging students appropriately. This analysis also gives him a way to measure different classes and students from one year to the next.

Venture Capitalist

The primary role of a venture capitalist is to provide some significant portion of the large startup costs for a new business. This is especially true for a business that might have a high risk that makes a firm unattractive to traditional bank financing. This risk is limited somewhat for the venture capitalist because unlike the bank manager, a venture capitalist is often an expert in their chosen industry or sector. That allows the venture capitalist to evaluate a business with a more expert opinion. In exchange for lending their capital, expertise, and connections to the entrepreneurs, these investors usually receive partial ownership of the business and a voice in key decision-making processes.

This can be a great way for new businesses or new products to raise funds that will help them through the obstacles in their way—like covering the cost of operating expenses (such as paying employees, rent for the building space, purchase of raw materials, and more) until the business is established and able to start generating a profit. At that point, the business might have the option to buy out the venture capitalist, or they might retain them as part of the management or as a member of the board of directors.

Failure is part of the game

In order to be aware of trends and hidden obstacles in the shifting business landscape, entrepreneurs and venture capitalists need to have an up-to-date picture of how many businesses are imploding each year (and why). If a firm wanted to run research to see what those failure rates looked like for business from zero to five years old, how large would the sample sizes need to be to get within:

a) 0.10 (10 percent) at 95 percent confidence?
b) 0.05 (5 percent) at 95 percent confidence?
c) 0.05 at 99 percent confidence?

To find the sample size, we start with the general formula:

$$n = \frac{p(1 - p)z^2}{m^2}$$

where:

n = Sample size.
p = Probability of success (here, 0.5 since success and failure seem equally likely).
z = Calculated number from a table.
m = Margin of error.

When we look up on the statistical tables, the z score for a 95 percent confidence interval is 1.96, so:

$$n = \frac{0.5(1 - 0.5)1.96^2}{0.10^2}$$

$$n = \frac{0.9604}{0.01} = 96.04 \approx 97$$

(We round up here to ensure we have a large enough sample size.)

I'll skip the middle one, and leave that for you test yourself if you're interested. The value you come up with should be 385.

Notice how big of a step up that is from 97 to 385. It's even worse for the final step, because in order to get up to 0.05 margin of error and 99 percent confidence, you would need to research a sample size of just a little over 650. That's a key takeaway from this example, though: increasing confidence and decreasing margin of error become more and more costly in time and resources for smaller and smaller gains in accuracy and confidence.

Bond bombshell

One of the ways businesses and investors raise money in a new venture is by selling bonds. These are like special loans. Consider an example where an investor buys an $8,000 bond from a new start-up, and the value of the bond is that the new company will pay $2,500 at the end of each year for five years. That seems like a pretty high rate of return, but the reason for that is that the investor is taking on risk. If we give the startup a 10 percent chance of failing within the first year, and then it goes down by 2 percent each year and levels out at 5 percent, then what are the chances now that the company will be around at the end of each of those five years to keep making payments on the bond?

Year	*Chance of Failure*	*Cumulative Chance Still in Business*
1	0.10	1 – 0.1 = 0.9
2	0.08	(1 – 0.08) × 0.9 = 0.828
3	0.06	(1 – 0.06) × 0.828 = 0.778
4	0.05	(1 – 0.05) × 0.778 = 0.739
5	0.05	(1 – 0.05) × 0.739 = 0.702

That means the investor who purchases the bond is taking a 30 percent chance the business will go under before it has made all its payments to him.

Estimating market control needed

As technologies mature, they support businesses that are valued by markets based on standard metrics. Let's take an example: A scientist has invented a radio that can communicate with any satellite in low Earth orbit. This is a newly emerging market. The radio is intelligent and cognitive, meaning that it can program itself to pick up any communication protocol it senses. The scientist claims that he can build a business, called BitRay, offering ground communications as a service to fledgling small satellite operators, in a manner similar to how wireless carriers offer subscriptions to mobile phone subscribers.

As a venture capitalist, you are evaluating an opportunity to invest in BitRay. As part of your diligence, you have observed that wireless carriers typically need to show at least 17 percent annual return on invested capital (IRR) to be viable. You have been asked to build a model to understand the level of market leverage and the number of small of satellites in orbit needed to hit the target IRR. Because this is a real-world example that involves a fair amount of complexity, the analysis would have to take a lot of different variables into account. A list of key variables needed would include:

- Fixed cost per ground station.
- Annual maintenance cost per ground station.
- Ground station lifetime.
- How much time a satellite spends over a ground station.
- How much BitRay can charge per minute of connectivity.
- Times a satellite orbits over a station per day.

Bonds

What in the world is a bond, actually? Mostly, bonds are a form of loan used by companies and governments. They seem to have been around since as old as recorded history and possibly before, with examples discovered in Mesopotamia dating back to 2400 BCE.

Why not just call it a loan, then? There are two primary differences between what we think of as a "normal" loan. (Keep in mind, though, in finance fields people have found out ways to invent all kinds of exceptions to most definitions we could offer.) The two primary differences are:

1. Bonds are designed in such a way that it is easy to buy and sell them. While you can sell normal loans, they are not designed for the sake of making them easy to buy and sell.
2. With bonds, there is rarely one person or institution that has loaned the money. Rather, in a way similar to stocks, each person buys a bond which is a small slice of the big loan that the company, city, or government is borrowing.

The fact that bonds are each a small slice from the bigger loan also helps make them much easier to buy and sell. In that way, they are also similar to stocks. However, with a stock you "own" a little slice of the company, which is not true with bonds. With a bond, the only thing you own is the rights to be repaid according to the terms of the bond.

- Serviceable lifespan of each satellite.
- Number of ground stations needed for global coverage.
- Amount BitRay needs to pay annually for backhaul communications.
- How much BitRay needs to pay annually for marketing expenses.
- How many satellites each ground station can support.

Different parts of that list would come into play at different stages of the project. To see how statistics start to factor in, let's grab a couple of closely related pieces and talk about them. The annual maintenance cost of a ground station is something we can average, but like the average student, that object doesn't really exist. Rather, we can look at a distribution, which would behave like a grade distribution on a test, roughly speaking. In the same way there is 10 percent chance, for example, of students getting 90 percent or higher on the test, a given ground station has an *x* percent chance of falling into the highest range of maintenance expenses. Right? Then the same idea would apply to the number likely to fall in the middle range of costs and then in the lower cost range. Those are closely connected to things like how long the ground station will last. Some percentage of the ground stations will die earlier than average, some later than average, and so on. Often the ones that die early would be the ones that might also have incurred the greatest amount of repair costs, but (since they died early) from which we have gotten the least value.

You can imagine how messy this set of relationships and probabilities starts to get. But wait, there's more! Investors don't always like companies where the value bounces around a lot, so a good model will figure out how to plan so that some of that bouncing around is smoothed out. For instance, the model might try to limit the probability that more than one ground station at a time might hit the end of its lifespan.

Similar calculations and adjustments would go on for other things, too: probabilities of launch failures, distribution of timelines. (What changes if a satellite launches late? How much chance is there it will launch late?) On top of the purely mechanical and engineer-y stuff, there is other business stuff that also has ranges and probability distributions. This includes things like market research, how many customers sign up, or how quickly new customers might demand the service. You get the idea. Those probabilities would then allow the investing firm to figure out whether BitRay will become profitable by 2020. They can come out with a carefully calculated, educated hypothesis for likelihood of the project collapsing at a given time. Based on all those calculations, they can figure out the growth and customer targets they would have to reach to get to 17 percent IRR. Feel free to take a minute and read back through to make sense of the idea. The goal here is not to seem heavy or confusing, just to give you an appreciation for how we can take those simple processes we've been talking about and stack them together to create a multi-kajillion dollar project.

Politicians

Politicians have responsibility for some portion of the functions of government at levels from the local town or village all the way up to the whole nation. Much of a politician's job is similar to that of a mid-level to upper-level manager in the business sector, with the key differences being that politicians manage public office, which is non-profit, and that they are generally in elected positions, which creates important differences between the decision-making process for them and for corporate managers.

Politicians' popularity

Even after elected, politicians are constantly working to build their approval rating among the public. There are several reasons for this, partly out of a desire to listen for what the people want, but also because having popular approval on your side of an issue or project lends momentum and gives the politician some power in getting other politicians to go along with a proposal.

Let's imagine we are looking at two public opinion polls about a month apart. When respondents were asked if they thought the president was doing a good job, there were some that answered the same way on both surveys, but there was also a smaller group whose answers changed. Some went from "yes" to "no," while others changed from "no" to "yes." Let's see how that plays out in terms of proportions.

	Second survey		
First Survey	**Yes**	**No**	***Total***
Yes	450	60	510
No	40	450	490
Total	490	510	1000

So, in the second survey, sixty people switched from "yes" to "no." Only forty people switched from "no" to "yes." An example of valuable information we could pull from this with a little statistics magic would be a way to determine if there is a meaningful difference in the proportion who said yes on the first survey and the proportion that said yes on the second survey.

In order to run the test for this, we need to assign values to the non-numerical options of yes and no. Let's have 1 be yes and 0 stand for no, and m_1 will stand for the mean of the first survey while m_2 stands for the mean of the second survey:

$$m_1 = \frac{(510 \times 1) + (490 \times 0)}{1000} = 0.51$$

$$m_2 = \frac{(490 \times 1) + (510 \times 0)}{1000} = 0.49$$

As we've learned, we can't always tell from looking at two means whether the difference between them is meaningful or not. (See that? Mean-ingful? Ha!) In cases like this, though, rather than the normal kind of equation, we can use a formula specifically for situations with yes/no answers. To do that, we need our two hypotheses:

H_0 = The null hypothesis, which stands for $m_1 = m_2$. This means there is *no* significant difference between the two means.

$H_a = m_1 \neq m_2$. This means there *is* a significant difference between the two.

The test statistic for this comes from the following equation:

$$z = \frac{no_2 - no_1}{\sqrt{no_2 + no_1}}$$

We take the number that switched from yes to no on the second survey and subtract the number who switched from no to yes. Then we divide that by the square root of their sums:

$$z = \frac{60 - 40}{\sqrt{60 + 40}} = \frac{20}{10} = 2$$

For a z value of 2, the key value is 0.046. That means there is only a 4.6 percent chance this difference is random. Since that is less than our confidence percentage of 0.05, we would reject the null hypothesis. Thus, yes, the means of the two surveys are significantly different.

Right to bear opinions

A large survey organization conducted a poll of just over one thousand American adults about whether the citizens thought there should be a law that bans possession of a handgun except for police and other similarly authorized personnel. Of the 1,012, there were 638 people who said we should not have that kind of law.

Does this represent a significant outcome? What is the margin of error and how confident are we that it is an accurate measurement?

With sample size n of 1,012, the proportion that said we should not ban handguns is $p = \frac{638}{1012}$, which is about 0.63. That means that the proportion of people who agreed with the proposed law is $1 - p = 0.37$.

With a confidence interval of 95 percent, our $Z_{\alpha/2}$ would equal 1.96 (taken from a table), so we can calculate the margin of error E:

$$E = Z_{\alpha/2} \times \sqrt{\frac{p(1-p)}{n}}$$

$$E = 1.96 \times \sqrt{\frac{0.63 \times 0.37}{1012}}$$

$$E = 0.02975 \approx 0.03$$

We then find the confidence interval by taking the original percentage plus and minus the margin of error:

$$0.63 \pm 0.03$$
$$= 0.60 \text{ lower bound}$$
$$= 0.66 \text{ upper bound}$$

That means we can say that we are 95 percent sure that between 60 and 66 percent of adult Americans think there should not be this law eliminating personal handgun ownership.

Grumbling over government

Suppose instead of evaluating the president's popularity, the politicians were trying to gauge the public's level of frustration with government in general—maybe because Congress has become gridlocked, maybe because internal discord within one of the parties has undermined its ability to generate consensus, or whatever it might be.

In a survey, respondents could choose either "things are fine," "I'm frustrated," or "I'm filled with rage against the (government) machine." The survey included 2,505 respondents, and the expected probability is that about 50 percent of people are "raging" at any given time. When the results came in, there were actually 56 percent who were raging against the machine. Our job, then, is to figure out if that difference is important or just a random, normal fluctuation. Our null hypothesis is that the difference was just a random fluctuation.

To compare the observed result against the expected one, we need to calculate the number expected to say they were angry. With the assumption that the normal level of angry respondents is 50 percent, then:

$$2505 \times 0.5 = 1253 \text{ people expected to be angry}$$

But in reality:

$$2505 \times 0.56 = 1402 \text{ people are angry,}$$
$$149 \text{ more than expected}$$

That leads us to calculate the standard error:

$$SE = \sqrt{\frac{p\,(1-p)}{n}}$$

where:

SE = Standard error.
p = Percentage of people expected to say "yes" to a question.
n = Number of people surveyed.

$$SE = \sqrt{\frac{0.5(1-0.5)}{2505}} = 0.00999$$

We can then put the pieces together to find a test statistic, z. This statistic subtracts the actual result from the expected one, and divides it by the standard error.

$$z = \frac{0.56 - 0.50}{0.00999} = 6.006$$

We would check that z value against a table and find that the probability of getting that value by chance is 0, which means we reject the null hypothesis. That means the survey result is truly different and can't be attributed to chance. People are more angry than politicians thought.

Do you kind of have a sense for what is happening in these cases? It's OK if you don't feel like you could do them—that's not our goal. The goal here is just for you to have a basic grasp of the ideas at work. If not, please take your time. Go back often and look at prior sections.

Counselor

Counseling is a career centered on caring for clients' mental health. Through a combination of research, client interaction, and continuing education, the counselor works with clients to help address issues and obstacles in the clients' mental and behavioral patterns. There will be significant variation depending on the specialty of the counselor's training, and this will range from alcohol and addiction counseling to school counseling, and many other choices in between.

Many of the functions of the profession might overlap with other careers like psychology and sociology, including ongoing research and academic conferences. But there are some central features that help to distinguish the field of counseling from others, including a focus on person-to-person therapy and working to inform and educate the public with respect to emotional health.

It isn't right to be left (out)

When we listen to people discuss what if feels like to be on the receiving end of social rejection—like being left out or excluded from a group—the language mirrors what we hear when people talk about physical abuse or injury. Counselors and psychologists run research projects to test what that means in terms of measurable evidence. One way to do that is to track differences in brain reactions in situations that imitate suddenly feeling excluded from some social activity. The resulting data might look like this:

Subject	Exclusion Rating	Brain Reaction
1	1.26	–0.055
2	1.85	–0.040
3	1.1	–0.026
4	2.5	–0.017
5	2.17	–0.017

Subject	Exclusion Rating	Brain Reaction
6	2.67	0.017
7	2.01	0.021
8	2.18	0.025
9	2.58	0.027
10	2.75	0.064
11	3.33	0.077
12	3.65	0.124

The "exclusion rating" comes from a brief questionnaire during the experiment. The brain reaction is an electronic measurement of brain activity in the area that registers physical pain responses.

When we run the crunch on these numbers, we find a strong positive correlation, where $r = 0.8782$. That means that social exclusion does cause a pain response similar to a physical injury.

A gift of wrapped happiness with each birthday

Sometimes we want to be able to use statistics to evaluate and analyze information that is not already being measured in some form of numbers. There are several ways to do that by assigning number values to different answers or rating scales. But we can also analyze certain kinds of answers without having to change them into numbers at all. We call those kinds of answers **qualitative**, which contrasts with quantitative (from "quantity," meaning numerical). The following example will help illustrate this idea.

In a study, 1,500 people were chosen at random and given a survey about how things are going in their lives. When the answers came back, researchers divided them up into age groups and then counted up their different responses. The underlying goal was to compare and contrast the younger generation with everyone older than them, so there were just two groups. The results looked like this:

	Age Group			
	18–25	26+	Total	Frequency
Very Happy	180	330	510	*34%*
Pretty Happy	378	435	813	*54.2%*
Not Too Happy	42	135	177	*11.8%*
Total	600	900	***1500***	
Frequency	*40%*	*60%*		

The goal here is to test to see if the two key variables are dependent or independent. That is, does being in one age group or the other have a significant impact on the likelihood of the answers given?

When working with qualitative data like this, the analysis uses something called a χ^2 test (chi-square test, where "chi" is pronounced with a *k* and long *i* sound). We set up two hypotheses:

$$H_0 : \text{The variables are independent}$$

$$H_a : \text{The variables are not independent}$$

We're testing whether or not two factors, age (*A*) and happiness (*B*), are independent. If they are, the probability of one item from each group being true is:

$$P(A) \times P(B)$$

In this case, the probability of being in the 18-25 group and of answering "very happy" would be:

$$0.4 \times 0.34 = 0.136$$

Then we use that to find what the expected number of people for this spot on the chart would be out of the total, so we take:

$$0.136 \times 1500 \text{ total people surveyed} = 204$$

If the variables were truly independent (that is, if age and happiness weren't related), we'd expected 204 to fill the upper-left cell of the table. If we repeat the same calculation for the other cells on the table, we would find the expected number of people in each cell should be:

	18–25	26+
Very Happy	204	306
Pretty Happy	325.2	487.8
Not Too Happy	70.8	106.2

Now we compare those expected values with what the researchers actually got back, and compare that to a critical value that we would look up on a table. (Note in a higher level stats course, you would spend time learning how those tables are constructed, but that is way beyond what we want to try to accomplish here.)

So, we have to calculate the number that we will check against the table. To do that, we add together the square of all the differences between observed values and expected values divided by the expected values. I know, it makes my brain hurt to read that too. Look at the equation and then look back at the words to see what it's showing:

$$\chi^2_{data} = \sum \frac{(O_i - E_i)^2}{E_i}$$

where:

O_i = Observed data at cell i (which changes).
E_i = The expected value at cell i.

Plugging in the numbers, the equation looks like this:

$$\chi^2_{data} = \frac{(180-204)^2}{204} + \frac{(330-306)^2}{306} + \frac{(378-325.2)^2}{325.2} + \frac{(435-487.8)^2}{487.8} + \frac{(42-70.8)^2}{70.8} + \frac{(135-106.2)^2}{106.2}$$

$$\chi^2_{data} = 38.5192$$

When we look on the reference chart for a chart which has this many rows and this many columns, the critical value against which we are testing our number is 5.991. Since 38.5192 is greater, the researchers would reject the H_0 hypothesis. There is evidence that a person's age has an effect on their answers to the questions about how things are going in their lives. This probably matches with what you might have guessed.

Making the conversation random and anonymous

Sometimes in working with group counseling scenarios, such as a group struggling with the same kind of mental illness or addiction, it can be very hard to get people to open up because they are afraid of being judged instead of being helped. There is a clever way that we can solve that issue, but still get people to share.

In a group session, a counselor could give each member a coin, a pen, and some paper. Each person would flip a coin. People who flipped and got heads would write down a story about their experiences with the difficult issue at hand. The people who got tails would just write some insightful comment about sports of some kind. Then the counselor collects the papers, and she could share some of the writings about the important topic, ignoring the ones about sports. This way, no one knows who is writing versus not writing about the sensitive issue, and even if the story is shared, no one will be able to guess who it belongs to.

Meteorologist

Most people think meteorologists make weather forecasts. That's a common and important task, but more broadly, meteorologists are scientists focused on the understanding of weather and climate. To assist them, they work with a wide variety of sensitive tools, instruments, and technology. The secondary function of this career is to translate their research into understandable insights and advice for the rest of the community.

Hurricanes are a good example of how events that seem to have a zillion variables can fit into a statistical pattern.

Rock you like a hurricane

Each year, a specialist formulates her prediction for the number of named storms that will develop for this year's hurricane season. Here is a table showing her predictions each year compared to the number of storms that actually arose.

Year	Predicted	Actual
1	8	6
2	11	8
3	9	7
4	12	19
5	10	13
6	11	7
7	10	14
8	14	12

Year	Predicted	Actual
9	12	14
10	12	15
11	11	12
12	14	16
13	14	14
14	15	27
15	17	9
16	17	14

If we made a line of best fit to compare the scientist's predictions with the actual number of storms, we would come up with $y = 1.803 + 0.157x$, where y is the actual number of storms and x is the predicted number. If our scientist predicts 16 named storms for the next season, what would we expect the actual value to be?

$$y = 1.803 + 0.9031x$$

$$y = 1.803 + 0.9031 \times 16$$

$$y = 1.803 + 14.45 = 16.25 \text{ storms}$$

Most of the time the scientist is pretty accurate in her predictions. What do you think the effect would be on the line of best fit if you were to rerun the calculations and omit the one outlying year, 14, where she predicted 15 but there ended up being 27 named storms?

Reservoirs

The US National Water and Climate Center keeps a careful record of the level of different lakes and reservoirs, measured as a percent of their total capacities. The capacity is the maximum amount of water a lake or reservoir could hold without flooding. Recording water levels is part of the process of tracking things like rainfall and seasonal variations. On a given day, scientists would run calculations on a random sample of lakes and reservoirs for three states in a region to compare and see if they had the same means for their water levels.

In the chart on the next page,

n = Number of items in the population .
m = The mean water level, as a percentage of capacity.
s^2 = The statistical variance.
s = The standard deviation (a measurement of how spread out the data is from the average).

After checking to make sure certain requirements are met for being able to run the calculations (like that the samples in one

State 1	% full
Source A	61
Source B	49
Source C	33
Source D	51
$n_1 = 4$	
$m_1 = 48.5$	
$s^2_1 = 134.3333$	
$s_1 = 11.5902$	

State 2	% full
Source A	80
Source B	83
Source C	77
Source D	94
Source E	84
Source F	68
$n_2 = 6$	
$m_2 = 81$	
$s^2_2 = 73.6$	
$s_2 = 8.5790$	

State 3	% full
Source A	35
Source B	46
Source C	55
Source D	53
Source E	46
$n_3 = 5$	
$m_3 = 47$	
$s^2_3 = 61.5$	
$s_3 = 7.8422$	

state were independent of the samples chosen from another state, that the standard deviations are not disproportionate, and so on) we can do an **analysis of variance (ANOVA)**.

The hypotheses being considered would be:

H_0 : The mean water levels are equal (have no significant difference).

H_a : The mean water levels are not all equal.

So, we calculate the value of the test statistic, F_{data}. The first step is to find the mean of all the populations put together. We add together all the values and divide by the total number of data points, which is fifteen. To make this faster, for each state we can do the calculation $n \times m$ (which is the average for that group).

Meet Carrie King, PhD

Here to offer a few thoughts is Carrie King, PhD, who was won numerous awards and honors in her work as a counselor (twenty-one years) and a professor of counseling (seven years).

Can you tell us a little about different ways that you use statistics as a counselor and in teaching other counselors?

In the counseling profession, we want to be effective in our work with helping others. Counselors do research and run experiments to find out new or better ways to help people achieve health and happiness. We also use statistics to evaluate if the new research is well done or if a specific testing instrument would be a good fit with a particular person.

Have there been things that have surprised you or opened your mind in a new way as a result of seeing what kind of picture or explanation data has revealed?

Part of my job is to advocate for others who are vulnerable or don't have power, so data that illustrates inequities—different rates of school discipline for students of different races, abilities, genders—is revealing and interesting to me. It points me in the direction I need to go in terms of further research and outreach. The same holds true if I see research that didn't include a diverse sample.

Do you think people would be surprised at how much math happens in your work?

Most people exploring counseling as a profession are surprised to know that they need to take a statistics course and a second course with lots of statistics just to get their master's degree. Counseling isn't just talking to people; we need to be able to read and understand research, as well as know how to conduct it.

I also run a graduate program training other counselors, so I use math to develop budgets, predict enrollment, plan for hiring of teachers, and to evaluate the effectiveness of our program.

As someone who spends her life understanding people's thinking patterns and processes, have you noticed that math affects yours in some ways?

Even when I'm not actively using math, the logic and problem-solving skills I learned by doing math are in constant use. It is as if learning and doing math is exercise for my brain. It has benefits to my learning and thinking even when there isn't a math problem in front of me.

$$\frac{4(48.5) + 6(81) + 5(47)}{15} = 61$$

Now we have to calculate how far each of the population means is from this new total mean. This calculation is known as the treatment sum of squares (*SSTR*).

$$SSTR = 4(48.5 - 61)^2 + 6(81 - 61)^2 + 5(47 - 61)^2$$
$$= 4{,}005$$

We then calculate the treatment mean square (*MSTR*). This measures how much the numbers bounce around in the sample means, weighted by the size of the sample. The *k* stands for the number of different populations. Here, we had samples from three different states, so:

$$MSTR = \frac{SSTR}{k - 1}$$

$$MSTR = \frac{4005}{3 - 1} = \frac{4005}{2} = 2{,}002.5$$

There are two more pieces to go. Next is the error sum of squares (*SSE*):

$$SSE = \sum(n - 1) \times s^2{}_n$$

$$SSE = [(4 - 1) \times 134.3333] + [(6 - 1) \times 73.6]$$
$$+ [(5 - 1) \times 61.5] = 1{,}017$$

This is telling us the sum of how far the individual data sets are off from their means. Then we use that to find the mean squared error (*MSE*):

$$MSE = \frac{SSE}{n - k} = \frac{1017}{15 - 3} = 84.75$$

This number is basically a measure of variability within each of the sample groups (how far off from the means they are compared to the size of the groups involved).

Now we have the pieces we need to find the F_{data} value:

$$F_{\text{data}} = \frac{MSTR}{MSE} = \frac{2002.5}{84.75} = 23.6$$

That number, then, can be checked against a table of values to see if it registers as significant. This table would tell us whether or not the differences were due to random fluctuations. In this case, the F_{data} value is significant, which means we would reject the null hypothesis (H_0) which said there was no real difference among the three different populations. In fact, the lakes in the different states have different average water levels.

This was a detailed example showing key steps in determining this result. It's okay if you don't quite understand all the terms here. It's important to understand that we determine how close a data set is to its mean. Then we can use statistical calculations to determine whether or not that mean is the same as others.

Meet Dave Young

Please meet Dave Young, a successful security and law enforcement trainer. With a background as a law enforcement officer, Dave has trained tens of thousands of officers and agents, as well as training other trainers in the last twenty years.

Thanks for joining us, Dave. Can you tell us what role statistics plays in your day-to-day career?

We use statistics to evaluate everything—the performance of our weapons, our strategy and tactics in the field, and the performance of our team members. I also use it to constantly evaluate our training programs to make sure we are cutting edge.

Are there ways you think most people would be surprised at how much math happens in your work?

You know, we train almost four thousand people every year. They become trainers who take what we teach them back to their squads and departments. That's a lot of people counting on us when you think about it, so we crunch numbers on everything. Current trends in criminal behavior, new equipment and resources, likely scenarios. When we recommend a piece of equipment or a strategy, you can bet we have done our homework.

Have there been things that have surprised you or opened your mind in a new way as a result of seeing what the data revealed?

I think the first time was when I really looked at the numbers and realized that over 98 percent of the people we train come back to us when it's time to renew their training qualifications. It was humbling to realize how much they believe in us to help them keep the people they watch over safe.

Is there a kind of math or kind of math problem you wish you were better at? Why?

Honestly, I wish I had pushed myself to take more stats. Each year, more and more police, military, and security personnel are using data to help them get better at their jobs, so the better we are at each element, the better we can help them become.

Any last thoughts you'd like to leave us with?

I'm telling you, math has saved my life, the lives of soldiers and officers under me, and the lives of people I've trained. End of story.

Thanks for your time and insights, Dave.

Monsoon

Sometimes the meteorologists are working to solve the puzzle in reverse. Consider, for instance, situations like in India where the monsoon season brings about 80 percent of their annual rainfall and is therefore critical to the success of agriculture. In studying data stretching back a little over a hundred years, researchers crunched numbers to find that the monsoons bring a mean rain level of 852 millimeters (33.5 inches) per year. If this has a standard deviation of 81.5 millimeters (3.2 inches), how do you think we could find the amount of rain that represents about 95 percent of all likely cases? (Think back to how we calculated confidence intervals and margins of error earlier in the book.) Similarly, how do you think we would calculate the percent chance that India might be hit by a drought year?

Law Enforcement and Security Training

This career cluster is focused on the training of civilian, government, and military personnel for law enforcement and security. While trainers need extensive familiarity with the detailed demands of law enforcement or corrections, that's not how they spend their time.

Some of the elements of trainers' work include reading to stay current on the most recent research available, constantly training themselves to maintain peak conditioning, and training classes of students in the techniques and strategies that will help them succeed in their public safety roles.

Security rotations

The four hundred employees at a private security agency are divided up at three different offices. Office 1 has 25 percent of the staff, Office 2 has 40 percent, and Office 3 holds the remaining 35 percent. In order to schedule their quarterly training updates and not cause too much disruption to any one facility by leaving it more shorthanded than necessary, the company doing the training needs to figure out a way to set up classes of twenty people each and have each class contain proportional representation from each of the three offices.

It might surprise you how much of solving real-world statistics problems involves first being able to set the problem up in a way that will allow for statistical calculations. Sometimes, the setup is harder than the calculations themselves!

Here, one basic idea for randomizing the group is actually to add a little bit of non-random framework. Since we need classes to be 40 percent from Office 2 (eight people), 25 percent from Office 1 (five people) and 35 percent from Office 3 (seven people), we make a set of cards that are labeled *o*-*e*, such as "1-15" or "3-4." Here, *o* stands for the office number, and *e* is the employee number. Then we'd randomly draw cards. First, we'd draw from the 1-*e* set of cards until we get five employees. Then we'd draw from the 2-*e* set of cards to get eight employees, and then we'd finish similarly with the third set of cards for the third office. By repeating this procedure, we'd eventually fill all classes with random employees in the correct proportions.

Digital detectives

Naturally, when we think about law enforcement teams keeping the world safe, we all tend to fall back on images from movies and television—running or car chases through the streets, maybe some explosions here and there. Not only are modern agencies using data to find and profile the people they are after, but in some interesting cases, the computer is actually the weapon of choice for collecting and analyzing the elements of the crime itself.

Consider the case of scandals that happen when a corporation is shuffling fake numbers around to deceive the public and the government while money is going out or coming in for inappropriate reasons. One of the ways the detectives working these cases can catch a break is by using the statistical probability that a particular

Digit	1	2	3	4	5	6	7	8	9
Probability	0.3	0.18	0.12	0.1	0.08	0.07	0.06	0.05	0.04

number will show up when numbers are actually happening from real stuff as opposed to numbers that have been artificially tweaked or randomly generated to try to cover up something.

Benford's law is one of those approaches. It describes how frequently the beginning digit of a real-life number occurs. The statistics it provides are shown in the table above.

Just to illustrate with an example, let's say that detectives are going through the accounting logs for campaign contributions from firms that just happened to receive some very **lucrative** contracts for construction work for the state. When the agent in charge runs the program to analyze statistical occurrences, she finds that the number 6 shows up as the first digit of entries at a rate of 0.113. That is much closer to the artificially random value of 1/9 = 0.111 than the realistic value of 0.07 that Benford's measuring stick would predict.

This would be one small clue, of course, and agents would be looking for more evidence than just that one discrepancy. One common mistake that goes against normal patterns is using 1 as the starting digit too often. Another mistake is having large sets of numbers with no repetitions or sequences—like never having amounts like 144.00 or 5.67. Those would occur with the same frequency as other numbers, so a large number of entries that never have doubles or runs would seem suspicious.

Data driving diva

A really exciting example of statistics being used to revolutionize the way law enforcement and sentencing is done is the research and resource development by Anne Milgram. She is the former attorney general of New Jersey and now a resident scholar at

New York University School of Law. Her current career focuses on work that she launched when she was the attorney general, where she was able to use data and analysis to drive all crime down by 26 percent and reduce violent crime by 41 percent.

By logging the data into the computer system and then crunching it, she and her team found surprising gaps and mismatches—like the high proportion of time being spent on criminal cases that had low impact and on criminals that had lower likelihood to be repeat offenders. At the heart of her resources is a tool that they developed to help judges with sentencing. This basically involved running correlation models, much like the ones we've been looking at in this book (but versions that have much more muscle than the ones here!). They ran these models on over nine hundred different factors involved in crime, repeat offenses, sentencing, and so on to come up with the nine most critical variables that judges can use, combined with the judges' education, training, and experience, to improve the impact and effectiveness of their sentencing decisions.

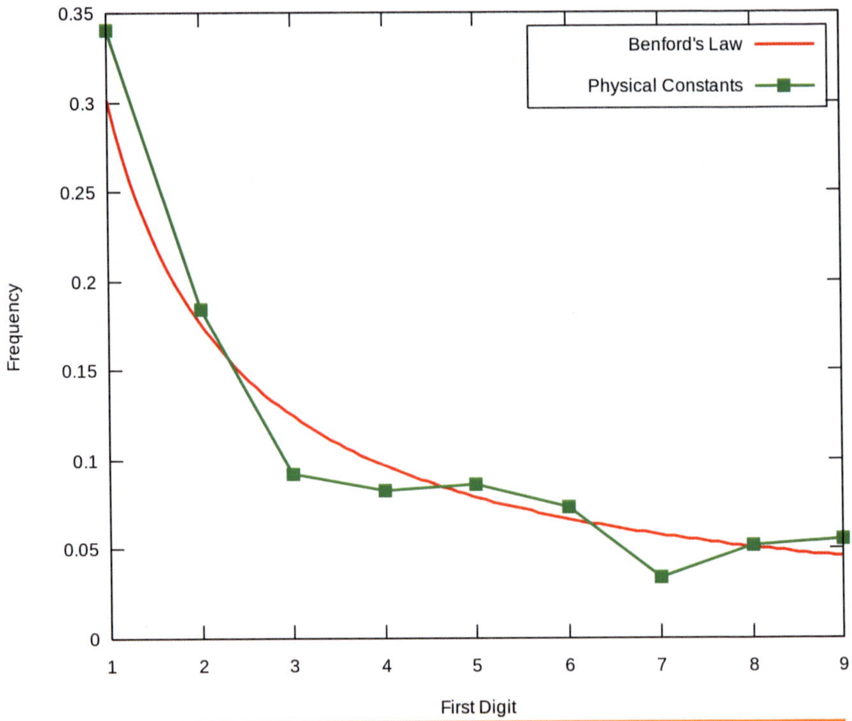

This graph compares the frequency of first digits of numbers in a data set as predicted by Benford's law to actual scientific data.

Writing Benford's law

The first person to notice the phenomenon that Benford describes was an American astronomer named Simon Newcomb back in 1881. He was using a book of logarithms, and happened to notice that the early pages of the book, the ones with numbers with 1 as the first digit, were considerably more beaten and bent. As he played with the idea, he noticed that the principle continued in diminishing amounts for other numbers. He came up with a formula, but then it was largely forgotten about.

In 1938, Frank Benford, an American physicist at General Electric, noticed the same pattern, but he decided to test it out on mountains of real-world sets of data. He pooled baseball stats, molecular weights, sizes of US city populations, data about rivers, and addresses. Even this weird assortment of data matched pretty well with the formula that Benford had cooked up.

The last piece of the puzzle came from Roger Pinkham, a New Jersey mathematician. In 1961, Pinkham became the first to try to figure out a mathematical proof for what was going on. Of particular coolness is that he proved Benford's law is scale invariant—it holds true for inches as much as for miles, for pennies as much as for dollars.

CONCLUSION

So here we are! You made it through without being eaten or poisoned or sucked into a black hole. Well done! The question is, did you look around on the way through? Did you notice the crystals sparkling on the walls of the cave? Did you stand at the edge of the rock outcropping and look over so you could have that slight dizzy feeling at the sheer height of it? I hope so.

What about stretching your mind? Did you imagine yourself looking out through other people's eyes to see the world differently? Every time we understand people a little better, then we understand ourselves a little better, too. Space and the rainforest and the amazing places we can travel all over this little blue water park we're riding on … those are nothing compared to the vast unexplored world inside the human mind. You get the idea.

But I want to come back around to the point with which we opened the book, and that is the idea that math is the tools and instruments that help us find our way. Think about this, because I think students often imagine that loving math means seeing the math equations the way other people see flowers or butterflies or stars at night. For some people that is true. For me it is somewhat true. If you are one of those people, great. However, I want to give you another possible way of seeing math and thinking about why it is exciting for some of us, and maybe why it is exciting for you, too.

Math is not the stars. Math is the telescope.

Like scuba gear, statistics and probability can allow us to go places (in our imagination, at least) that would otherwise be off limits.

Let me say that again:

Math is not the stars. Math is the telescope.

It is night vision goggles so you can see the amazing rainforest critters that only come out after the sun goes down. It is the scuba tank that allows you to see the coral reef close up. Breathing the oxygen in will help to keep your brain from getting foggy, or your sight from getting dim. Seriously. Your brain will work clearer when you have been using it for math. Don't take my word for it, though. Run the experiment for yourself—but for it to be a fair experiment, you have to be willing to accept the evidence, right? Do some extra math every day for a couple weeks. Do it for thirty minutes. Maybe that means trying the challenge problem on your homework, but maybe not. That's fine. Go on a website and read about some new math discovery. Go online and play any number of math games. You can even do some math puzzles and brainteasers. No pressure. No grades or tests for this part. It's fun. It's tinkering to see if you can get your brain to do this cool thing where it lights up a little.

Have fun. Have fun playing with math. Use math to help you have fun doing other stuff!

Glossary

altimeter An instrument that measures your distance above sea level, usually based on pressure.

analysis of variance (ANOVA) A method that compares the means of two (or more) different groups of data to see if they are significantly different.

assumptions Items taken as true without the intent or the ability to justify them on further grounds.

celestial Having to do with sky, space, or objects in space.

commodities Things that are bought and sold. Often refers to raw materials or other inputs to a production process.

confidence interval A range of values within which the overall mean for a population should lie. The level of confidence is expressed as a percentage; for example, a 95 percent confidence interval indicates 95 percent certainty that the mean falls within the stated range.

correlation How closely two things are related or dependent on each other.

cryptanalysis The process of analyzing and breaking a code.

data Facts or statistics collected for the primary purpose of research.

demography The study of statistical data relating to a population or smaller subgroups within it.

electrons The small, negatively charged particles orbiting the nucleus of an atom.

encapsulate To close up or wrap up something in a summary or within a broad concept.

fluctuation The rising and falling of something; when values for a variable move up and down.

frequency The rate at which something repeats in a given period of time.

homicide When one person kills another; often a result of defense, neglect, or in commission of a crime.

hybrid A new version created with features from two or more older versions of something.

hypothesis A claim supported by reasoning that is being tested by experimentation.

lucrative Describing something that is very profitable.

mean Another term for what we normally call the average. It is the result of adding all the data values together and then dividing by the total number of values.

median The middle number in a data set that is arranged from lowest to highest values. If there is an even number of data points, then we add the two numbers in the middle and divide by two.

mortality rate Given a certain age, disease, or other condition, the likelihood that a person will die.

outlier A data point that is outside the normal range.

percentile On a chart of performance measures, it is the number that tells what percentage of other data points are below a certain one.

phenomena Things that happen; different occurrences.

plausible Seemingly true or reasonable.

qualitative Measuring or fitting something into categories by some means other than a number.

quantitative Studies or research utilizing measurable numerical values.

randomize To make unpredictable by taking away patterns or trends.

regression The measurement of the strength of relation between two variables.

standard deviation A measurement showing how much the numbers in a data set are spread out. A low number would mean they are grouped really closely together—each one a small distance away from the mean.

standardized tests Tests in which everyone answers the same questions and statistical analysis is employed to make sure the tests are evaluated in a way that is consistent across large groups to allow for valid comparisons.

supernova An exploding star which suddenly expands in brightness and size, often unexpectedly.

tangible Something you can touch.

variable An element or part of a formula or equation that can stand in for different possible values.

Further Reading

Books

Bluman, Allan G. *Probability Demystified: Hard Stuff Made Easy*. New York: McGraw-Hill Education, 2012.

Gowers, Timothy, June Barrow-Green, and Imre Leader, eds. *The Princeton Companion to Mathematics*. Princeton, NJ: Princeton University Press, 2003.

Jackson, Tom, ed. *Mathematics: An Illustrated History of Numbers*. New York: Shelter Harbor Press, 2012.

Levine, David M, David F. Stephan. *Even You Can Learn Statistics: A Guide for Everyone Who Has Ever Been Afraid of Statistics*. Upper Saddle River, NJ: Pearson, 2010.

Pickover, Clifford A. *The Math Book: From Pythagoras to the 57th Dimension, 250 Milestones in the History of Mathematics*. New York: Sterling, 2009.

Websites

Carnegie-Mellon—The Data and Story Library
lib.stat.cmu.edu/DASL

Carnegie Mellon University has created a nice resource that is searchable by concept or topic to find stories that illustrate key ideas and to use stored data sets to populate some of your statistics problems and examples.

Cool Math
www.coolmath.com/precalculus-review-calculus-intro

This is a good website for a wide range of topics. Of particular note is the colorful layout for a good basic review of pre-calculus material.

Dartmouth—Chance
www.dartmouth.edu/~chance

Dartmouth developed this program to provide resources and materials for running a course in quantitative literacy. Even if you don't work through the lessons, some of the videos are helpful in getting your brain around some of the central concepts.

Drexel University Math Forum
mathforum.org/dr.math

This site offers a variety of basic explanations of concepts and examples. One of its strengths is that the topics are broken out by subjects *and* by level in school (elementary, middle, high school, college, and beyond). They do a helpful job of working through example problems and also allow you to search by topic.

Khan Academy
www.khanacademy.org

This is an excellent online learning website, and the learn-as-you-go format allows you to click on only the short instructional videos you need as you work your way through the course. There are courses at a wide range of skill levels.

NYU Department of Mathematics—Courant Institute
cims.nyu.edu/~kiryl/elementary_statistics.html

This is a nice, clean, visual layout of a thorough variety of topics in statistics. It connects to other sections which review key ideas from algebra, geometry, trigonometry, etc.

***Plus* Magazine**
plus.maths.org/content

This online magazine is run under the Millennium Mathematics Project at Cambridge University. It offers interesting short articles on interesting problems in math (and science). The problems are complex, but they are broken down in a way that is easy to follow.

Bibliography

Agresti, Alan, Christine Franklin. *Statistics: The Art and Science of Learning from Data*. 3rd ed. Upper Saddle River, NJ: Pearson, 2013.

Bellos, Alex. *The Grapes of Math*. New York: Simon and Schuster, 2014.

Berlinghoff, William P., Kerry E. Grant, and Dale Skrien. *A Mathematical Sampler: Topics for Liberal Arts*. Lanham, MD: Ardsley House Publishers, 2001.

Byers, William. *How Mathematicians Think: Using Ambiguity, Contradiction, and Paradox to Create Mathematics*. Princeton, NJ: Princeton University Press, 2007.

Clawson, Calvin C. *Mathematical Mysteries: The Beauty and Magic of Numbers*. New York: Plenum Press, 1996.

Devlin, Keith. *Mathematics: The Science of Patterns: The Search for Order in Life, Mind and the Universe*. New York: Henry Holt, 2003.

Ellenberg, Jordan. *How Not to Be Wrong: The Power of Mathematical Thinking*. New York: Penguin, 2014.

Freeman, W. H., Brendan Cady, ed. *For All Practical Purposes: Mathematical Literacy in Today's World*. New York: W. H. Freeman and Company, 2006.

Frenkel, Edward. *Love & Math: The Heart of Hidden Reality*. New York: Basic Books, 2013.

Gravetter, Frederick J., Larry B. Wallnau. *Statistics for the Behavioral Sciences*. 9th ed. Belmont, CA: Wadsworth, 2013.

Jackson, Tom, ed. *Mathematics: An Illustrated History of Numbers*. New York: Shelter Harbor Press, 2012.

Larose, Daniel T. *Discovering Statistics*. New York: W. H. Freeman & Company, 2010.

Mankiewicz, Richard. *The Story of Mathematics*. Princeton, NJ: Princeton University Press, 2000.

McLeish, John. *Numbers: The History of Numbers and How They Shape Our Lives*. New York: Fawcett Columbine, 1991.

Moore, David S. *The Basic Practice of Statistics*. 5th ed. New York: W. H. Freeman & Company, 2010.

Oakley, Barber. *A Mind for Numbers: How to Excel at Math and Science (Even if You Flunked Algebra)*. New York: Tarcher, 2014.

Paulos, John Allen. *Beyond Numeracy: Ruminations of a Numbers Man*. New York: Alfred A. Knopf, 1991.

Peterson, Ivars. *Islands of Truth: A Mathematical Mystery Cruise*. New York: W. H. Freeman and Company, 1990.

———. *The Mathematical Tourist: Snapshots of Modern Mathematics*. New York: W. H. Freeman and Company, 1988.

Pickover, Clifford A. *The Math Book: From Pythagoras to the 57th Dimension, 250 Milestones in the History of Mathematics*. New York: Sterling, 2009.

Rooney, Anne. *The Story of Mathematics: From Creating the Pyramids to Exploring Infinity*. London: Arcturus, 2015.

Sardar, Ziauddin, Jerry Ravetz, and Borin Van Loon. *Introducing Mathematics*. Cambridge, UK: Icon Books, 1999.

Stewart, Ian. *Nature's Numbers: The Unreal Reality of Mathematics*. New York: Basic Books, 1995.

Strogatz, Steven. *The Joy of X: A Guided Tour of Math, from One to Infinity*. Boston: Houghton-Mifflin, 2012.

Weiss, Neil A. *Elementary Statistics*. 8th ed. Boston: Addison-Wesley, 2012.

Index

Page numbers in **boldface** are illustrations. Entries in **boldface** are glossary terms.

About the Author

Erik Richardson is an award-winning teacher from Milwaukee, where he has taught and tutored math up to the college level over the last ten years. He has done graduate work in math, economics, and the philosophy of math, and he uses all three in his work as a business consultant with corporations and small businesses. He is a member of the Kappa Mu Epsilon math honor society, and some of his work applying math to different kinds of problems has shown up at conferences, in magazines, and even in a few pieces of published poetry. As the director of Every Einstein (everyeinstein.org), he works actively to get math and science resources into the hands of teachers and students all over the country.